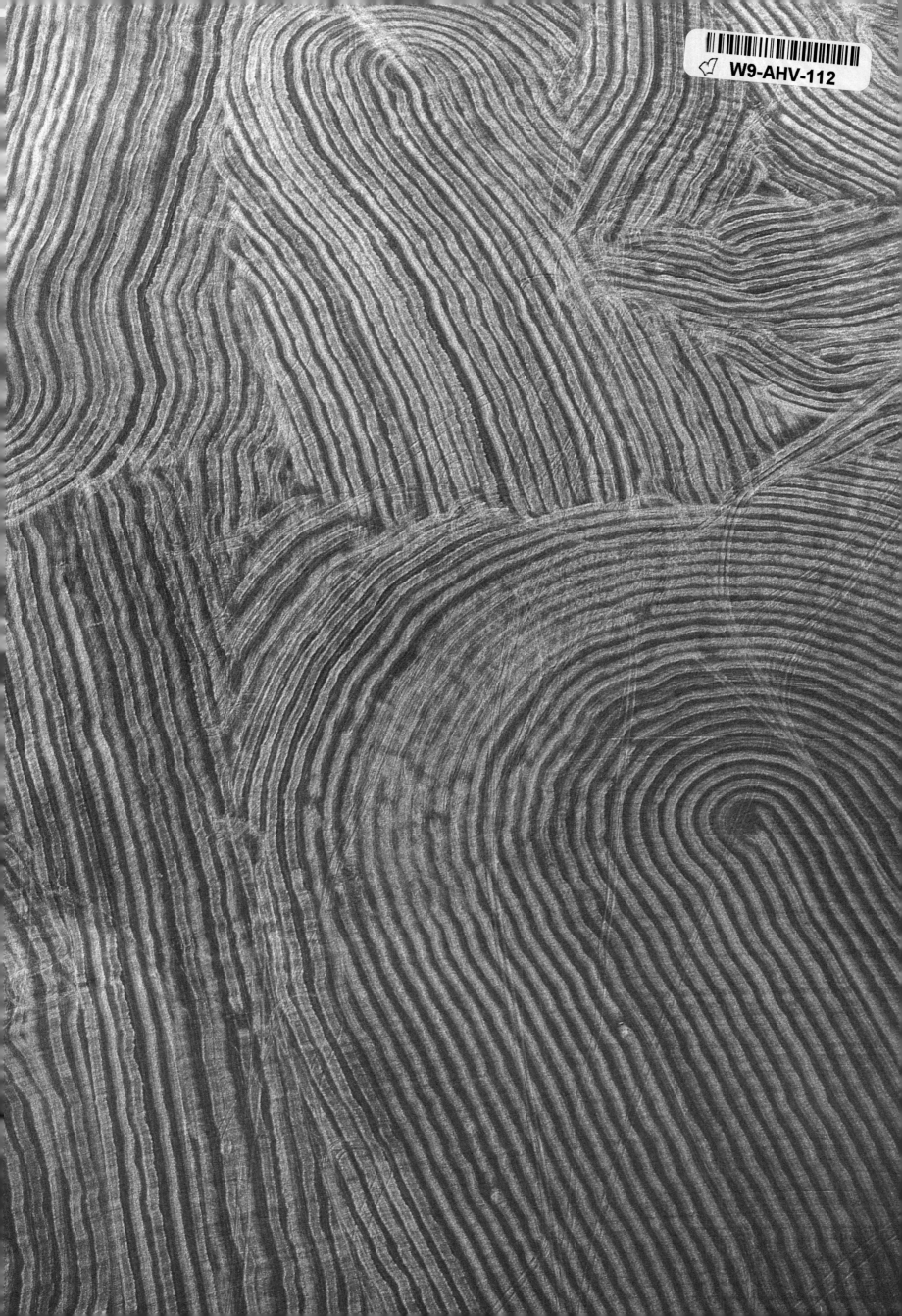

Georg Gerster

Amber Waves of Grain

America's Farmlands from Above

GEORG GERSTER

AMBER WAVES OF GRAIN

America's Farmlands from Above

Text and Captions by
JOYCE DIAMANTI

Foreword by
GARRISON KEILLOR

HARPER · WELDON · OWEN

IN ASSOCIATION WITH THE AMERICAN FARMLAND TRUST

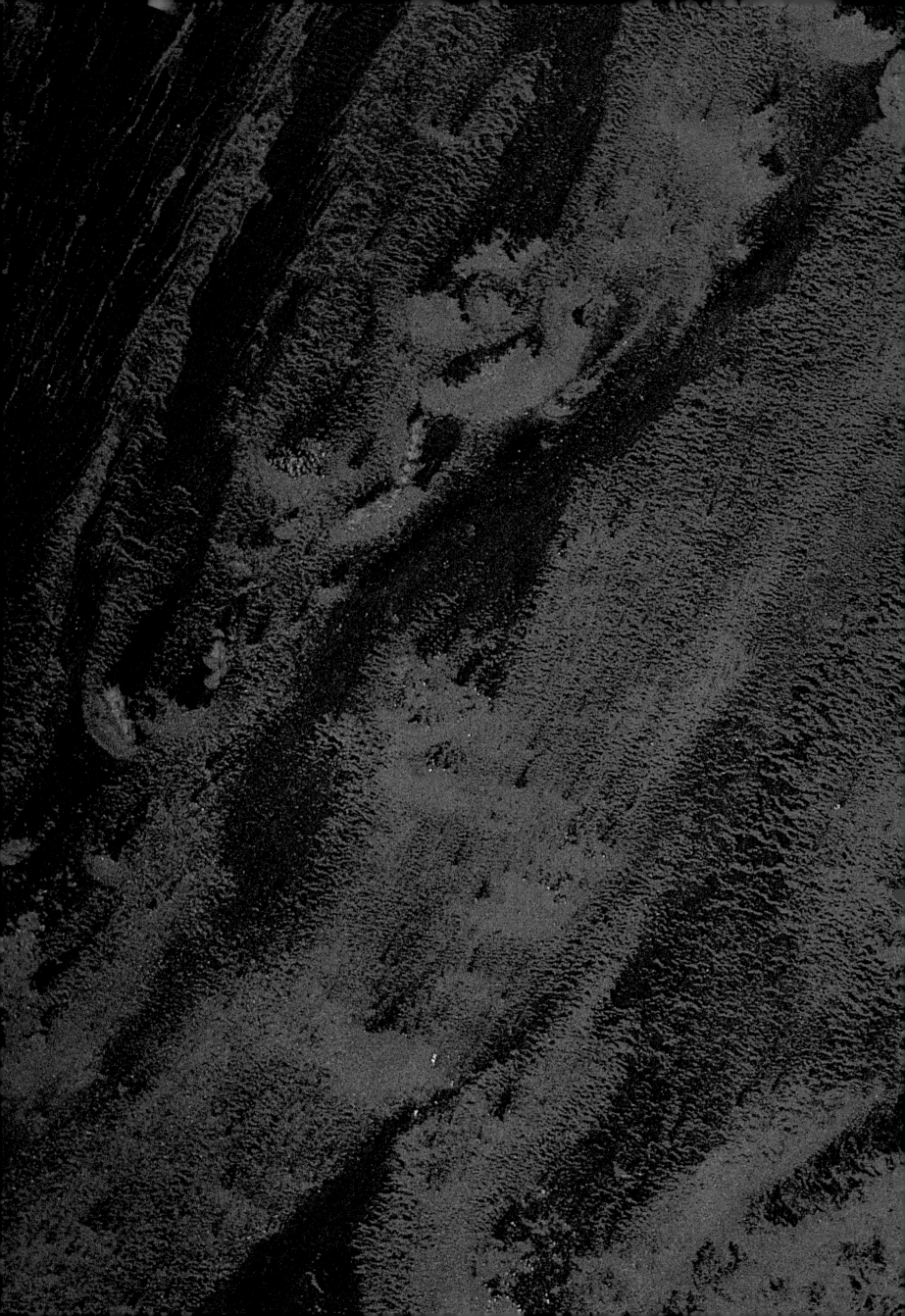

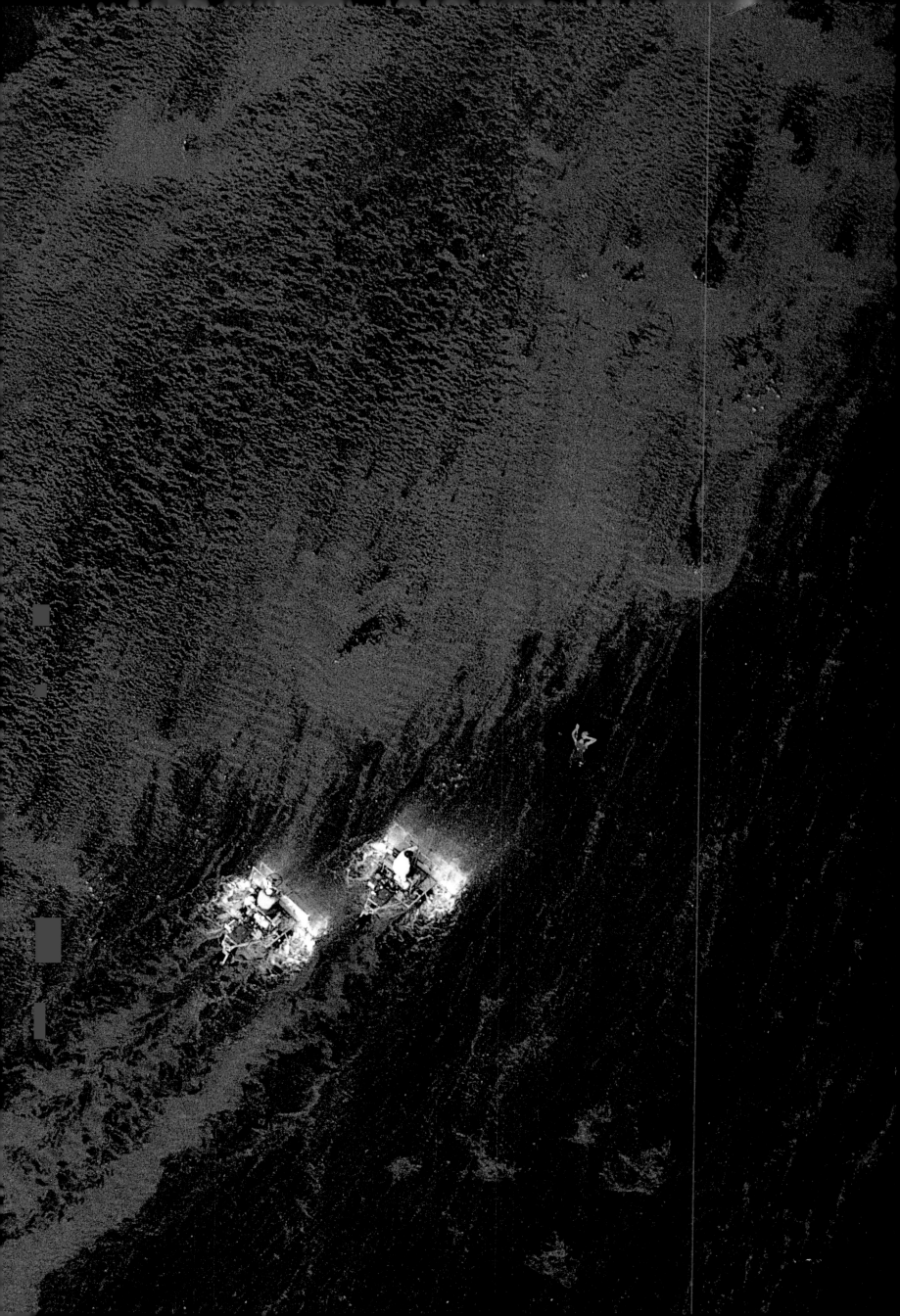

First published in 1990
by Harper Weldon Owen, New York

Conceived and produced by Weldon Owen Inc., 90 Gold Street,
San Francisco, CA 94133 Tel (415) 291-0100, Fax (415) 291-8841

Designed by Constance H. Phelps

Weldon Owen Inc.

Publishing Manager: Stuart Laurence
Managing Editors: Beverley Barnes, Sheena Coupe
Editor: Daphne Rawling
Typography: Warren Penney
Assembly: Ingrid Padina
Maps: Mark Seidler
President: John Owen
Chairman: Kevin Weldon
Consultant: American Farmland Trust

HarperCollins Publishers

Publisher: William M. Shinker
Senior Editor: Susan Friedland

Library of Congress Cataloging-in-Publication Data

Gerster, Georg. 1928–
    Amber waves of grain: America's farmlands from above/Georg
  Gerster; texts and captions by Joyce Diamanti; foreword by
  Garrison Keillor. — 1st ed.
        p.   cm.
    "In association with American Farmland Trust."
    ISBN 0-06-016463-8
    1. Farms — United States. 2. Agriculture — United States.
  3. Farms — United States — Pictorial works. 4. Agriculture — United
  States — Pictorial works. 5. United States — Description and
  travel — 1981 – — Views. I. Diamanti, Joyce. II. American Farmland
  Trust. III. Title.
  S441. G47   1990   90-4553
  779'. 963' 0973 — dc20

Typeset by Amazing Faces, Sydney, Australia
Produced in Hong Kong by Mandarin Offset

Printed in Hong Kong

A Weldon Owen Production

Note on the maps
Photographs throughout this book are coded, above their captions,
with colored dots which are reproduced on the accompanying maps.
Each dot on a map indicates the state in which the photographs coded
with the same color were taken.

Endsheets: Corrugated windrows loop across the landscape in Nez
Perce County, Idaho. The ridges of hay – or perhaps bluegrass grown
for seed, or peas or lentils – were made by a swather to ensure fast and
even ripening before the harvest is gathered in by a baler or a combine.

Pages 2–3: Contoured strips of hay, corn, and new-sown grain
emblazon a dairy farm in southeastern Pennsylvania, where, harvest
after harvest, the land has kept its promise of plenty for nearly
three centuries.

Page 4: Rows of ripe corn swerve around a pile of stones wrested from
rocky Vermont fields tilled by the same family for generations; in New
England, farming has been likened to "getting crops out of granite."

Pages 6–7: Churning the water like eggbeaters, two water reels,
guided through a flooded cranberry bog by a hip-booted signalman,
shake ripe berries loose for harvesting near Plymouth, Massachusetts.

Right: Driving a red tractor, artist Stan Herd etches "Sunflower Still
Life" out of a 20-acre clover field near Eudora, Kansas. His crop art will
gain texture and hue as soybeans and sunflowers sprout and mature.

Pages 10–11: A crumbling limestone outcropping near Scottsbluff,
Nebraska, divides gullied rangeland from terraced cropland laid out
with alternating contour strips of wheat and fallow to conserve soil
and moisture.

Page 12: Green and yellow tongues of cropland ooze out like spilled
paint among scattered ponderosa pines, near Flora, in northeastern
Oregon. The tortuous outlines of the field are determined by the
availability of soil, a thin and precious resource in this scabland region.

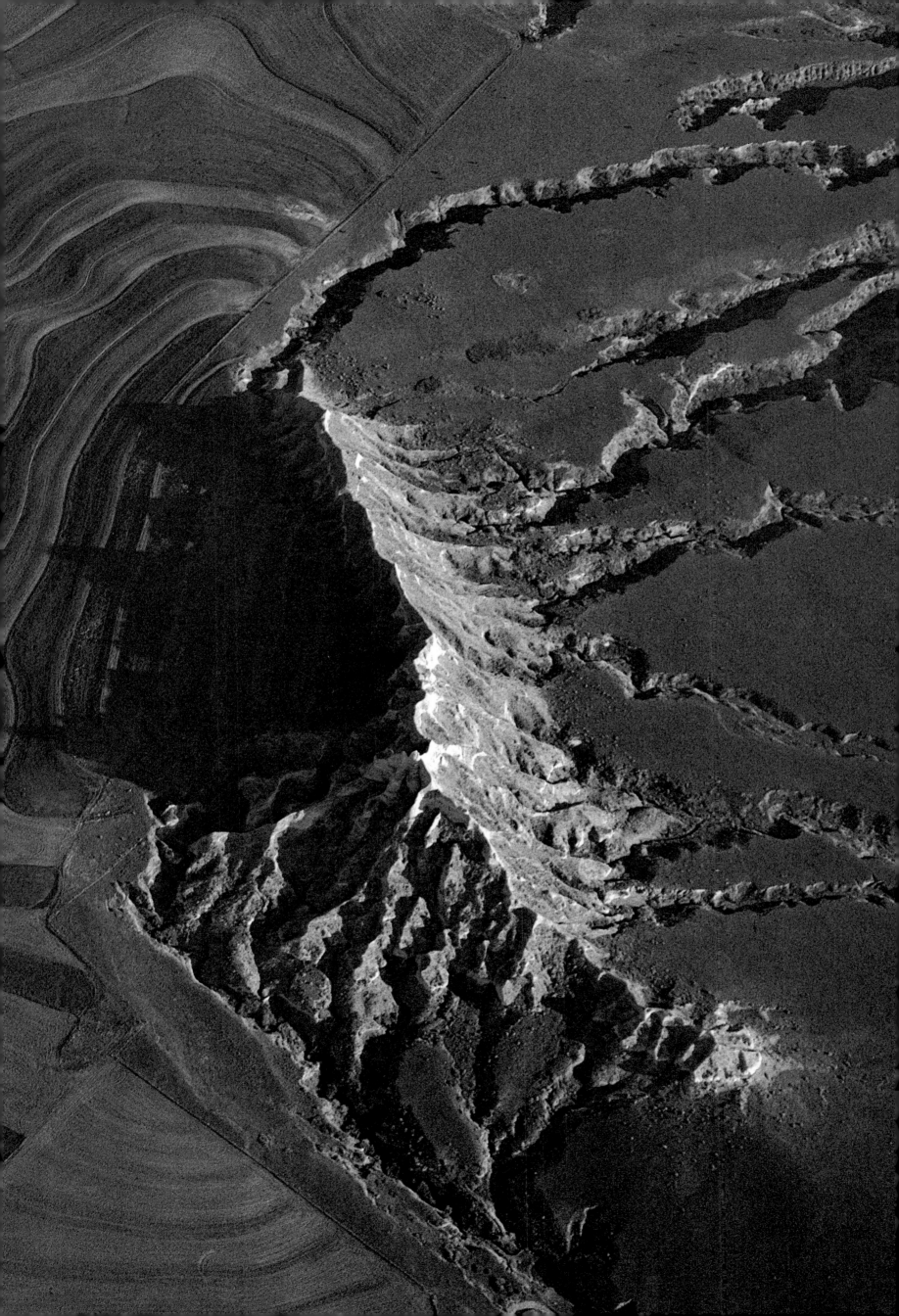

FOREWORD

In the kitchen of Grace and Gordon Linke's house near Chaffee, North Dakota, hangs a magnificent aerial photograph of their farm, taken by a local pilot in summertime from directly overhead, a lush green paradise picture. It was given to Uncle Gordon by the local conservation district in recognition of his faithful work in soil conservation, says a little brass plaque beneath, and the picture clearly shows the long green shelter belts that he was so proud of. You can see every outbuilding, the roof of the house shaded by elms, the Chevy pickup, even a slight shadow cast by the mailbox.

Uncle Gordon hung the picture in such a prominent place, on the wall next to the mud room, because it gave him pleasure to look at it. A six-hundred-acre spread, hogs, corn, oats, flax, sunflowers, the Lord's own view of it as He smiled down on the Linke family. Many times a day, on his way out to chores and back for breakfast, out, back for lunch, back out and home for supper and one last trip out to check on things, as Gordon took off his barn boots and his jacket and hung his cap on the hook, he glanced at the photograph of his farm. It was a spiritual icon, pointed out to every visitor. Here in Grace's kitchen, snug and warm, smelling faintly of yeast and roast pork and lemon detergent, next to the mud room smelling faintly of Gordon and Red River Valley loam and pig manure, you could see where they had spent their lives and what they worked so hard for. You could also feel a child's metaphysical thrill: here am I looking at the picture of the farmhouse where I am standing looking at the picture of the farmhouse with me standing in it.

Gordon died a few years ago and I think of him again as I look at Georg Gerster's magnificent low flight over America. Looking at these pictures is like what I as a boy imagined dead people could enjoy as they floated around, revisiting their old neighborhoods, hearing distant piano music, remembering what life was like and how sweet corn tasted.

In picture after picture, hovering overhead, your imagination descends and your feet touch down ghost-like in that gorgeous field of alfalfa encircled by fall maples and oaks; you walk the sea of yellow sunflowers; you pass invisibly through the handsome homestead, the barn and sheds and garage and milkhouse and coops and pens and cribs, and into the woodlot, like the spirit of Gordon or any other farmer, perpetually checking the property, seeing how things are going.

Most Americans are not far removed from the farm, and our driving desire to leave the farm and move to the city doesn't cut our ties to the place, not nearly. Old farm kids sit in glass-and-steel splendor and maneuver huge corporations and eat at La Place de la Cuisine and sail their boats beyond sight of land, but a few photographs can put you right back where you came from, back in your mother's kitchen. The evening of a hot July day, you've just finished washing dishes after supper and you walk out the kitchen door into the yard. Twilight gathers in the grass. In the cool of the evening, the farm seems to gather itself together. You walk down the two-rut road, out of the yard, out of the windbreak, out into the open.

On the plains, our ancestors planted trees close by their houses so as to break up the view, which can be too long for a person to live with — when you see so far, you only see trouble, all the things that can go wrong and so few people to set it all right. But in the evening, you lie on your back and look a billion miles into the sky filled with billions of stars and you can feel yourself float free of the earth, free of this farm, this year, this self, sailing out into the dark and brilliant future.

In Georg Gerster's photographs, we're lying on our backs in the clouds, looking down at the land, a radically different view than from the tractor seat or the Interstate. In these pictures, we're already floating. On our backs in the clouds, we see constellations of rock, mythical topography, ancient shores, some shapes chosen and made by us, others providential, and the fragile shapes of farms, so earnest, naked, so eloquent of the hopes and labors of men and women. We look down to bless them.

Garrison Keillor

GARRISON KEILLOR

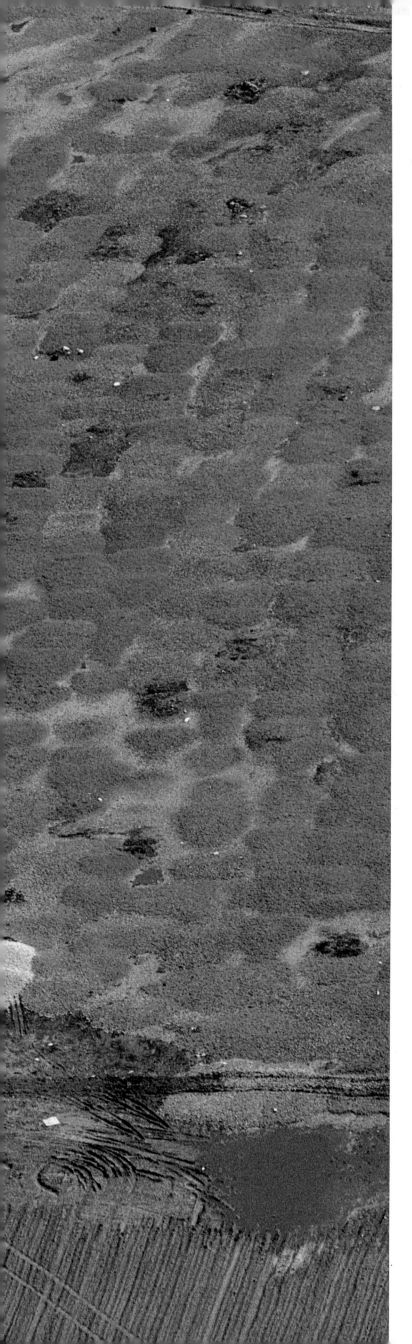

CONTENTS

Left: Truckloads of red and green tomatoes, orange and yellow squash – "graded out" because they failed to meet standards of size, shape, or color or other criteria of perfection – decompose at a disposal site near Homestead, Florida.

Overleaf: Migrant crews wielding knives harvest lettuce near Las Cruces, New Mexico, as boxes are constructed in the field. Almost all of the nation's six-billion-pound lettuce crop is hand-harvested.

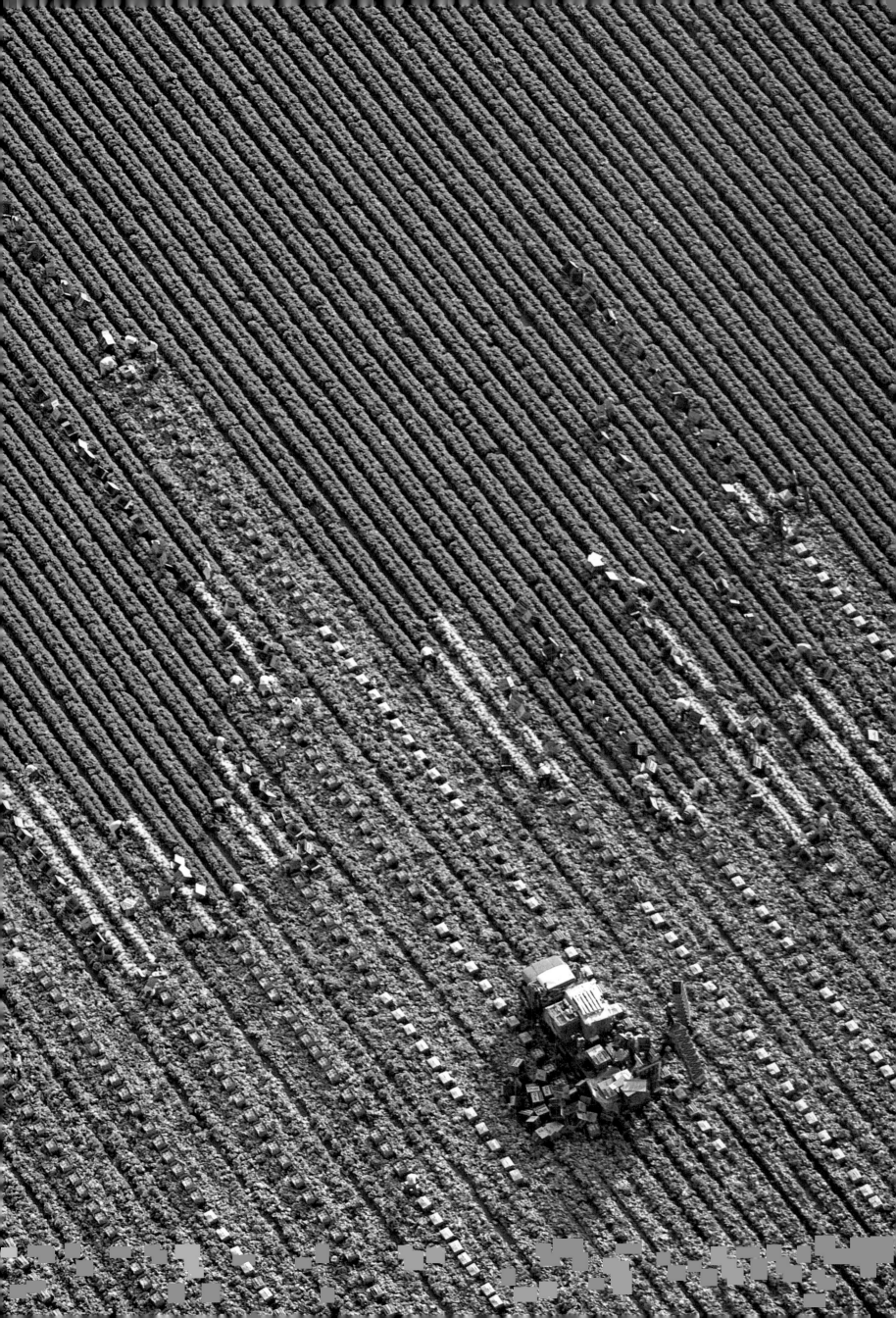

PREFACE

Georg Gerster

This book is about the art in and the art of U.S. farming as seen from an eagle's eye. It aims, on the one hand, to hold up a mirror in which American farmers can recognize themselves with pride and, on the other hand, to entice city-dwellers to a new appreciation of their provider. At the outset of the project, I was satisfied with finding mere beauty: winging over farmlands in all fifty States I serendipitously chased after the eye-dazzling. It took a thousand hours in low-flying small aircraft and two years – if I add up the ground time related to these photo flights – to sharpen my sight for the issues lurking beneath the merely beautiful. I can only hope that the images of the book, chosen from one hundred thousand transparencies, reflect this learning process.

All over the world farmers draw with plow, harrow, and harvesting combine, and paint with the colors of their crops. As land artists they have no equal, and the palette and patterns of American agriculture, in particular, seem inexhaustible. While coaxing bounty from the earth to feed the United States and other nations, farmers coincidentally offer up a visual feast for the eyes of an airborne viewer. To such a one the agrarian landscapes between the Hudson and the Sacramento appear as a vast open-air museum, with hundreds of thousands of tableaux on display, most of them a square mile in area and framed by bordering roads. Some of these exhibits rival the mystery of prehistoric ground drawings; others conjure up the tumultuous abstractions of modern canvases. And all of them incessantly change with the seasons, being blessed with both transient and timeless splendor.

I cannot hide the fact that many of the pilots who took me up only reluctantly shared my rapture. Professional caution urged them to forever be on the lookout for a landing site should anything go wrong. (Nothing ever did.) Their concern over where they might safely set down their craft tempered their delight. Corn on the contour is a visual favorite of mine. "The worst!" exclaimed Gary Blessing, as he piloted me at harvest time over lush fields of corn near Shenandoah, Iowa. "Corn stalks are tall and sturdy, and the contours indicate a terrain anything but flat. Corn is a no-no even at spring planting time, for then the ridges between the furrows stand forbiddingly high." A wheat field, I learned from Gary, is the next best thing to a meadow for an emergency landing. Soybeans range somewhere between wheat and corn; sorghum is the same as corn to a pilot – shorter but equally uninviting. Clearly, one man's signal is another man's noise.

Speaking of divergent perspectives, American farmers, needless to say, do not harbor artistic intent, they are land artists without really trying – usually. A few exceptions of planned field graphics just prove the rule. Neighboring counties where large-scale farmers fly to work or into town in their own planes have had contests in land art as well as in crop yields. In the late 1930s, a farmer in Bent County, Colorado, plowed a mosaic of circular strips into the native grass on a full square-mile section: a helper, acting as a center pivot, reined him with a rope as he was going round. He had no justification for his creation other than that he liked doing it. For the past two decades this particular field has been conventionally block farmed, with total disregard for the former artistic layout, but flying over it I could still discern ghostly circles, like old script underneath newer lettering on a reused parchment. In the 1960s a Wisconsin patriot patterned his farm after the American flag. And in 1988 an Iowa farmer planted a section of corn in the shape of 60-year-old Mickey Mouse. The awareness that airborne viewers are looking down on a farmer's

Left: Like giant moth wings, boldly patterned fields of winter wheat spread over the rolling hills of southeastern Washington. Contour strip-cropping reduced erosion on this farm by more than 70 percent. The dark strips have lain fallow for a season to store up moisture; the pale strips are stubble left as cover after the August harvest; the golden centers are grain still to be cut.

19

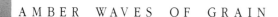

ALASKA

Juneau

work inspires not only aesthetic impulses. Instead of sending telegrams to Washington, American farmers sometimes inscribe their fields. I AM BROKE MR. REAGAN! wrote an Illinois farmer, in one of the longer and more polite messages crying out to the heavens. The texts prompted in the 1980s by the farm crisis more often consisted of four-letter words. At times, indignation even speaks without words. During World War II a farmer in Gray County, Kansas, was driven to despair by the noise of Flying Fortresses carrying out training flights from two nearby air bases. His exasperation obviously got the better of his judgement, for he plowed a huge swastika in his field. The provocation did not fail to command the attention of the pilots. And the protesting farmer was lucky that their only retaliation was to bombard him with even more noise.

Farmers can, of course, almost always come up with sober and pedestrian reasons for their land art; they are responding to the requirements of terrain and soil, to the climate, their machinery, or the market. But this does not preclude them from adding a dash of the irrational on occasion – even farmers sometimes run away from their problems (and their spouses). After bad dust storms struck eastern New Mexico in February 1977, soil scientists investigating the source of the problem found that some farmers who knew better had frivolously plowed their furrows at right angles to the direction recommended for halting wind erosion – "out of boredom and disgust with the monotony of my work," said one. And another farmer admitted that he had senselessly plowed up a field protected by stubble after having quarreled with his wife – "I had to let off steam somehow."

But enough of artistic fancy and fits of temper. The average American farmer is a man of reason, and his unintentional art derives solely from measures aimed at higher or at least sustained yields. I found it gratifying that the most stunning productions in this open-air museum originate in the resolve not to mine but to husband the soil. Stewardship of the soil, born of the near catastrophe of the "Dirty Thirties," has enlivened the writing on the land with the graphic vocabulary of conservation farming. Perhaps no farming area in the U.S. is visually more exciting than Lancaster, Berks and York Counties in Pennsylvania; none, I am told, is more productive.

When asked by American friends about the rationale of my work I once suggested: "The French gave you the Statue of Liberty. Why shouldn't a Swiss offer you a new perception of your land?" – and immediately regretted the smugness of my reply. I admit to misgivings. How to justify my lofty search for that gossamer web named beauty when on the ground many farmers struggled with harsh realities, gripped by drought, squeezed by rising costs and sagging prices, facing foreclosure?

I was delivered from doubt in Lewiston, Idaho, where I saw how beauty can spur beneficial action. A conservation technician, David Hein, had pinned up one of my photographs in his office – the aerial view that opens this preface. It shows a farm near Pomeroy, Washington, with an exemplary layout to combat erosion. The photograph was noticed by Alex Schaub, a farmer who had stubbornly refused help from the Soil Conversation Service to save his ailing 2000-acre farm. He studied the picture for a while, then announced that David and his service could have a free hand, "if you can make my farm look as fantastic as that."

David could and did. Within two years he had given the Schaub farm the zebra look of the model. Stripcropping on the contour, together with crop rotation, reduced soil loss from

WASHINGTON

Olympia

Salem

OREGON

IDAHO

Boise

Hel

Carson City

Sacramento

NEVADA

Salt Lake Ci

UTAH

CALIFORNIA

ARIZONA

Phoenix

Honolulu

HAWAII

the previous high of 156 tons per year per acre to a tolerable four tons. "You've saved the country millions of tons of the best agricultural soil," David assured me as we flew together over "my" farm, now rescued from perdition by soil surgery.

A compliment could hardly be weightier. But the ultimate accolade awaited me in the most light-minded of places, Las Vegas.

Donn Owens had piloted me on one of the last sorties for this project. After our return to the North Las Vegas airport he drove me back to my hotel. Dusk fell; the glitz and razzamatazz of gambler's paradise beckoned with myriad lights. Luminous but utterly devoid of numinosity, the city was an improbable backdrop for what Donn, born of an Oglala Sioux mother, was to tell me. While a student at Washington State University in Pullman, he had come across some of my aerial views of the Palouse country in eastern Washington. "I was profoundly moved by them", Donn recalled. "My heritage, I guess. Your pictures are imbued with an almost Indian sensitivity for the land." Native Americans, he said, bemoan the broken bonds with the land. "Whenever the ancient guardian spirits of the land – invoked by whatever means, be it even a photograph – reappear and speak to us, we resonate."

Donn's words suffused me with the satisfaction that comes with a wonderfully soft touch-down: a long quest above had attained its goal below.

A BEAUTIFUL,
BOUNTIFUL LAND

O beautiful for spacious skies,
For amber waves of grain,
For purple mountain majesties
Above the fruited plain!
America! America!
God shed His grace on thee . . .

Beautiful from time immemorial, when God alone was the beholder, America has become the most bountiful land on earth, where in the late twentieth century a farmer's labor yields, on average, enough food and fiber to meet the needs of 117 people. Annually Americans consume, per capita, 171 pounds of fresh fruits and vegetables, 210 pounds of red meat and poultry, 168 pounds of sugar and sweeteners, and 293 pounds of milk, cream, butter, cheese and ice cream. Yet we spend less than 15 percent of our disposable income on food, compared to more than 30 percent spent by consumers in some industrialized nations and more than 60 percent in some developing countries.

Today the United States is home to less than 3/10 of 1 percent of the world's farmworkers. But they produce 14 percent of the world's wheat, 17 percent of its cotton, 23 percent of its beef, 25 percent of its oranges, 31 percent of its poultry, 47 percent of its corn, and 59 percent of its soybeans. The harvest of about one acre in three goes abroad – more American land than Japanese land is devoted to putting food on the tables of Japan – and our farm exports surpass those of any other country.

How did America achieve such abundance? We must of course give thanks for nature's twin blessings of fertile soil and a favorable climate. But crops and livestock do not grow, bear fruit, and multiply on their own; it is the hand of man that made the continent a cornucopia. Americans of many professions and persuasions – pilgrims, pioneers, presidents; lawmakers, lawbreakers; scientists, inventors, entrepreneurs – have made unique contributions to our agricultural success. Above all, credit is due to the men, women, and children who over the centuries have tilled the soil, in short, to the dirt farmers who, by their will, wit, and toil, realized the primeval promise of the land we now call America.

When the first migrants reached North America from Asia, much of the continent lay locked in the glacial grip of the last ice age. The newcomers survived by hunting and fishing, gathering seasonal nuts and berries, and foraging for edible seeds and tubers. As early as 11,000 BC cave dwellers in what is now Idaho had domesticated the dog. But as the climate warmed and agriculture developed in the New World, from about 7,000 BC, the only plants to be domesticated by inhabitants of the temperate swath that sweeps across the continent's midsection, and now comprises the forty-eight contiguous states, were the sunflower and the tepary bean, a drought-resistant native of the Southwest. And the only other animal they tamed was the turkey. Save for a few recently domesticated species, all the other crops and livestock that make up our present agricultural bounty are, like American agriculturalists themselves, immigrants or the offspring of immigrants.

Maize, or corn, by far our biggest crop today in terms of both acreage planted and cash value, was first to arrive, along with squash and common beans, spreading north from Mesoamerica in the first millennium AD. By the time European explorers began to probe North America's bays and inlets, Indians were cultivating this trio of crops in village plots and open fields of up to 200 acres. Reconnoitering our southern coasts in 1584, Captain Arthur Barlowe reported that the natives were friendly and the corn they gave him was

Left: Cornfields converge at the Indian town of Secota on the Carolina coast, in an engraving made by Theodore de Bry in 1590, after a watercolor by colonist John White. In field H is "corne newly sprong," in G "greene corne," and in F "rype corne" guarded by a watchman in a shelter who wards off feathered raiders with "cryes and noyse." Other cultivated crops include pumpkins overflowing a broad border (I) and tobacco and sunflowers set out in garden plots (E).

Pages 22–23: Pear trees in bloom spangle the Wenatchee River valley in central Washington. Snaking through the orchard, an irrigation ditch feeds sprinklers which supplement scanty summer rainfall.

Pages 24–25: Shimmering in the glow of early morning, wheatfields on the Nez Perce Indian Reservation in northern Idaho evoke those of Hamlin Garland's nineteenth-century boyhood: "Deep as the breast of a man, wide as the sea . . . a meeting place of winds and of sunlight – our fields ran to the world's end."

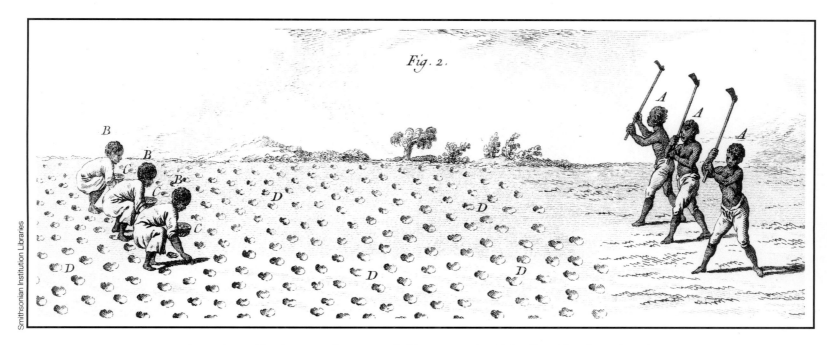

Indigo cultivated by slaves was the most profitable crop in colonial Carolina. After the spring rains, men moving backwards across a field made planting holes with hoes, while crouching women advanced, dropping five or six seeds in each hole then covering them with a sweep of the foot. Many planters imported African slaves who had been trained in the West Indies, the site of these engravings.

"white, tempting, and of excellent flavor. It is harvested three times in five months," he marveled. "The soil is deep, sweet, healthy, and the most fruitful in the world."

Thomas Hariot, surveyor of England's first colony, on Roanoke Island, expanded on the virtues of maize: "We made malt from the grain . . . and brewed as good an ale of it as could be desired." As to kernels grown for each kernel sown, he informed investors that "the grain increases on a marvelous scale – a thousand times, fifteen hundred, and in some cases two thousand fold." In the wholesome climate, he claimed, an acre yielded at least 200 bushels of corn and beans as well as pumpkins, gourds, and sunflowers set out among the corn hills.

When the *Mayflower* made landfall in the winter of 1620, however, the Pilgrim Father William Bradford found Cape Cod to be a "hideous and desolate wilderness, full of wild beasts and wild men." Yet the colony the Pilgrims founded at Plymouth flourished, in no small part because the Indians generously shared their agricultural know-how and the Pilgrims were quick to adopt it. "Many ways hath their advice and endeavour been advantageous to us," a colonist recorded, "they being our first instructors for the planting of their Indian corn, by teaching us to cull out the finest seed, to observe the fittest season, to keep distance for holes and fit measure for hills."

Most of the early European immigrants knew little about farming. Predominantly town and village folk, they were merchants and craftsmen by trade – tailors, carpenters, printers, cobblers. Their ranks included common laborers and highborn gentry, as well as paupers, prostitutes, and convicted criminals who had chosen the unknown of the New World over Old World prisons. They came seeking fortune in the form of gold or furs, seeking religious or political freedom, adventure or romance. Many, victims of hard times, were simply seeking a second chance. Almost all turned to farming to survive.

Boundless land beckoned, but before the colonists could cultivate it they had to clear it of dense virgin forest. A handbook for "the Unexperienced Planters of New England, or Anywhere" explained how to "spoil the woods" by the Indian method of girdling. Instead of chopping down trees and grubbing up stumps, a farmer was advised to strip a band of bark from each tree's trunk. Within a year or so the branches would die off, and the farmer could plant the ground exposed to the sun; later, he could burn off the dead trunks and leave the stumps to rot. Still, most settlers managed to clear no more than an acre or two a year, so at best it took half a century to wrest a hundred acres of cropland from the wilderness.

Gaining title to the land was another challenge. And again, the settlers took a distinctly American approach. Not peasants bound by tradition, they refused to be held thrall to tenancy or quitrent, thus preventing a feudal system of land tenure from taking hold in the New World. Though the British crown granted rights over immense areas to colonization companies and a few individuals, their schemes proved unprofitable or unenforceable in the vastness of America.

In the South, companies that tried to use indentured labor to bring their territory under cultivation were soon obliged to grant headrights of 50 acres or more to lure settlers, and by the eighteenth century land

When the indigo blossomed in early July, it was ripe for the reaper's hook. Harvesters stooped low to
cut the stalks close to the ground; they had to keep their sickles sharp and slice cleanly, so the bushlike
indigo plants would grow back for two more cuttings. Here a woman bundles stalks as a fieldhand
carries a sheaf off to be made into dye for the European textile market.

acquisition had passed to purchase. The thrifty Pilgrims of New England saved their profits from fur and
timber, and within two decades bought out the British-based Plymouth Company. They parcelled out farm
plots to the faithful, but each village kept a common for pasture and a green for recreation and militia drill.

In the middle colonies, William Penn's domain sprawled over 47 million acres, making this unlikely
Quaker the second largest landowner in the world in the 1680s. An ardent promoter of the family farm,
Penn tried to distribute the land by selling it for as little as two pounds sterling per hundred acres and
granting generous headrights to settlers who paid their own passage. Penn's heirs offered the ultimate bargain, "walking purchase" of as much land as a man could pace off (or run around) in a day and a half. Even
so, in 1726 more than 100,000 squatters were tilling land in Pennsylvania to which they had no title. Recog-
nizing the right of preemption, the government let these farmers buy cheap and pay later.

By the eighteenth century, then, many strands had come together to create the pattern of independent
farmers that has marked our nation's agriculture. On the eve of the Revolution some 90 percent of the
American people, including many of our founding fathers, lived and worked on farms, and most farmers
owned their own land. Today only 2 percent of the population live on farms, but about 85 percent of American farm operators still own, in whole or in part, the land they work. In 1782 J. Hector St. John de
Crèvecoeur wrote, in *Letters from an American Farmer,* "We are a people of cultivators . . . all animated
with the spirit of an industry which is unfettered and unrestrained, because each person works for himself." Americans, he explained, reap the rewards of their labor "without any part being claimed, either by a
despotic prince, a rich abbot, or a mighty lord," and farmers are "anxious to get as much as they can,
because what they get is their own."

American agriculture has been commercial from the start, for beyond what was needed for subsistence,
settlers strove to bring in a surplus they could sell at a profit. Their goal was not survival but advancement.
And to this end American farmers have shunned neither hard work nor high risks – trying new crops, new
techniques, new technology; buying, borrowing, always scrambling to adapt to a changing economy; forever breaking new ground, finding new problems, seeking new solutions – in pursuit of ever greater abundance and a bigger marketable surplus. Today America's agribusiness, which includes transporting, processing, and retailing farm output as well as growing food and fiber, accounts for about 17 percent of the
gross national product. With assets of nearly $1 trillion, it is the world's largest commercial industry.

In colonial America, northern farmers produced grain, flax, and livestock for domestic and foreign markets. The "Bread Colonies," notably New York and Pennsylvania, exported wheat and flour to the West
Indies. But Europe absorbed the bulk of the agricultural surplus, and the most important export crops
originated in the South.

The first big cash crop was a New World native – tobacco. In 1616 Virginia exported 2,500 pounds of the
lucrative leaf, and farmers were so smitten with "tobacco fever" that the governor had to order them to plant
a minimal two acres of corn per person. Around 1674 tobacco planters imported the first African slaves to

work their fields. Within a century three-quarters of Maryland and Virginia landowners had become slaveholders, and tobacco exports had soared to more than 100 million pounds per annum.

Thus a third agricultural tradition, from Africa, merged with Indian and European traditions to create a unique legacy for American farmers. Indians had bred high-yielding native crops and developed tools and techniques suited to the American terrain and climate. Europeans brought Old World plants and animals, Iron Age technology, and a fund of empirical knowledge. Africans contributed sorghum, spices, and perhaps rice, and well-honed skills in hoe-and-sickle agriculture.

In 1700 blacks made up about a tenth of the colonial population; by independence they constituted about a fifth of the population, numbering half a million, preponderantly slaves in the southern states. The plantation pattern of agriculture grew out of exploitation of this labor pool and monoculture of the great commercial crops that thrived in the South – tobacco, rice, indigo, sugar and cotton.

Rice flourished in inland swamps and man-made paddies in the lower South and became the region's major export in the eighteenth century. Indigo for the British textile industry was a profitable companion crop. Both crops were harvested by sickle, and their processing required much handwork, but differing seasonal demands assured efficient year-round use of slave labor. Indigo lost its market during the Revolution. Rice continued to be important, but in the nineteenth century it lost its supremacy to King Cotton.

In 1793 Eli Whitney, a Yankee tutor sojourning in Georgia, took up the challenge of local planters to invent a machine that could separate the seeds quickly and cleanly from the fiber of short-staple cotton. A tinkerer since boyhood, he took only ten days to build a model of the gin that would make a historic breakthrough in cotton production. Within ten years cotton overtook tobacco exports; by 1820 it accounted for more than half of all farm exports; and by the late 1850s cotton surpassed in value all other exports combined, agricultural and manufactured. But despite such technological advances, stoop labor remained the foundation of cotton production. Planting called for teams of three, one slave to hoe, one to drop seed, and one to cover the seed with soil. And harvesting premium export cotton, without leaves and dirt, required hand-picking.

Increasing European demand spurred the spread of cotton culture into the interior. In 1811 the advent of steamboats on the rivers that drain America's heartland ushered in an era of fast, cheap inland transport, giving impetus to the westward movement of northern as well as southern farmers. Within a generation more than 1,000 stern- and sidewheelers were churning our western waters, carrying some ten million tons of freight annually, while sleek clipper ships provided access to foreign markets.

In colonial times a few farmers, defying a British ban on western settlement, had trickled through the gaps of the Appalachians. By 1850 the agricultural frontier had surged 500 miles westward and leaped the Mississippi. In the next five decades the tide swept across 1,000 miles of prairies and plains and crested the Rockies to merge with a backwash from California. By 1900 American farmers, encouraged by federal policies, had brought the continent under cultivation from coast to coast.

Very early, legislation with the dual aim of promoting settlement and raising revenue set the pattern for the disposal of public lands the Republic had fallen heir to and new territories it later acquired. The Land Ordinance of 1785 established the rectangular survey system, which has left its imprint on much of the landscape. Large blocks, or townships, six miles on a side, were divided into 36 square-mile sections comprising 640 acres each. The land was then sold at auction, with a minimum bid of a dollar an acre. The Northwest Ordinance of 1787 provided for territorial government; by prohibiting laws of entail and primogeniture, which fetter land transfer, it helped assure a flexible market for American farmland.

The Act of 1796 and subsequent laws facilitated land acquisition by small farmers. Local land officers were opened to conduct auctions. Although the minimum bid per acre was raised to two dollars, credit was granted. And the minimum parcel put on the block was gradually reduced from a 640-acre section – about eight times more land than one family could farm – to a 160-acre quarter section, the usual unit of sale, and eventually to a 40-acre quarter-quarter section. Preemption acts legalized settlement before purchase, long a de facto feature of the American frontier.

The promise of law, order, and land lured pioneers to treeless western expanses where breaking virgin prairie presented an even greater challenge than clearing virgin forest. The thickly interwoven roots of perennial grasses, deeply embedded in sticky lime-rich soil, defied conventional plows. Massive wooden sod breakers were fitted with wrought-iron shares that could be sharpened in the field. Pulled by teams of four horses or as many as seven yoke of oxen, the largest plows could turn a 30-inch furrow. "The tough vinelike roots of the bluestems and prairie clovers twanged with a thousand ringing sounds as each step of the horses pulled the sharp blade forward through the sod of ten thousand years," wrote Hamlin Garland, chronicler of the northern prairie. "As the soil was turned, a black ribbon reinforced with the living foundation of the prairie thudded into the trench made by the preceding round of the plow. Bee nests were upturned, mice scurried from their ruined homes, and raucous gulls swarmed behind, picking up worms,

AMERICAN PROGRESS, a color lithograph after an 1872 painting by John Gast, depicts Indians yielding to waves of westward expansion from the urban East. The trapper, the cowboy, and miners are followed by farmers with the "sacred plow"; the covered wagon is succeeded by the stagecoach then the railroad. Presiding over all, the spirit of Progress carries a "School Book" and strings telegraph wire, bringing civilization to the vast, amorphous continent.

insects and other small creatures evicted from the sanctuary of their grassland homes."

Early settlers labored to break eight acres of sod in a season. Three freshly turned furrows provided a protective ring against prairie wildfires; the following year the farmer would plant these rows and advance into the surrounding grassland by plowing a new firebreak. To prepare the broken ground before planting his first corn crop, he used a cast-iron plow. But sticky clods of earth and decaying sod mired the plow as often as every few feet, forcing the farmer to scrape off the moldboard before he could go on. This second plowing, in the wake of sod busting, proved so arduous that some speculated that America's prairie lands – now the richest agricultural area on earth – would have to be abandoned.

But plowmakers continued to experiment, and in 1837 John Deere designed a plow with a steel share smoothly welded to a wrought-iron moldboard. So cleanly did it scour, it became known as the "singing plow." By 1857 Deere was selling 10,000 prairie plows annually, and others were producing similar models.

In the first half of the nineteenth century the national domain expanded more than threefold through the acquisition of territory from France, Spain, Great Britain, and Mexico and the annexation of Texas. The need to secure these new lands and make them productive overrode the desire to raise revenue, and in 1862 Congress passed the Homestead Act, which offered land free to settlers. By farming or residing on a claim for five years, any American citizen, or immigrant who had filed intention papers, could gain title to 160 acres.

As railroads stretched west after 1850, they opened up new farming regions. To subsidize construction of transcontinental lines, states and the federal government made land grants to the railroads in zones as wide as 40 miles on either side of the right-of-way. In their efforts to sell this land to raise capital, the railroads posted agents at US ports of entry and even in Europe to entice immigrants. In 1874 the Sante Fe sold 60,000 acres in Kansas to a congregation of Mennonites from Russia. To tout the potential of the Dakotas, the Northern Pacific backed a seasoned wheat farmer from Minnesota with ample capital and eighteen sections of bottomland in the Red River Valley. He hired gangs of workers and ordered carloads of the latest equipment. So spectacular was his success — well over 100 percent profit on cost of cultivation — that within five years the 300-mile-long valley, bought up by eastern syndicates, was filled with "bonanza"

Courtesy of the Library of Congress

"This brigade of horse artillery sweeps by in echelon – in close order, reaper following reaper," wrote an awe-struck reporter, describing the conquest of the Dakotas by mechanized agriculture in the late 1800s. These self-binding reapers saved both time and labor. Paddles on the left of each machine cut the wheat, while the mechanism on the right bound the stalks and, amid a clatter of cogs and iron arms, tossed the sheaves out, ready for threshing.

A steam tractor drives a threshing machine on a bonanza farm in the Red River Valley in 1878. As a belt transmits power to the thresher, a large crew unloads wheat from field wagons to feed its ravenous maw, helps bag the grain – typically 600–800 bushels a day – and piles the straw in stacks. Some steam engines burned straw, but most were fueled by coal or wood. Barrels of water for filling the boiler also came in handy for dousing fires set by sparks.

farms, some as large as 100,000 acres. And the Northern Pacific reaped a second harvest, this time in freight charges for hauling the bounty to market.

A similar land-grant program financed higher education for the sons and daughters of farmers and other "ordinary citizens". Under the Morrill Act of 1862, land-grant colleges were established in every state, offering instruction in agriculture, engineering, and home economics as well as in the traditional disciplines. Subsequent legislation funded experiment stations and created the Extension Service to bring practical information directly to farmers. This investment in education and research would pay dividends as American agriculture changed course from expansion to productivity.

In the last three decades of the nineteenth century, 430,000,000 acres were settled and 225,000,000 brought under cultivation, more than in the previous three centuries of American history. A "Permanent Indian Frontier," beyond which Native Americans would roam forever free from white encroachment, had proven ephemeral. It was repeatedly pushed westward, until in 1887 the Dawes Act provided for the dissolution of tribal reserves, setting the stage for a series of epic land runs. The curtain went up at noon on April 22, 1889, when shots rang out and 50,000 to 100,000 land-hungry "Boomers" — mounted on horses, mules, and even bicycles; spilling from wagons; clinging to the roofs and cowcatchers of overflowing trains — surged into Oklahoma District to pound their claim stakes deep in the heart of Indian Territory. An untold number of "Sooners" had slipped past patrols and jumped the gun. By nightfall 1,920,000 acres had been "settled." In 1893 the last large tract of tribal land was thrown open to homesteaders. By 1900 the best agricultural land in America was occupied; further expansion would be into marginal areas.

The force that drove our farm frontiers westward was not population pressure but commercial enterprise geared to the growing urban markets of the United States and Europe. Coinciding with expansion, the Industrial Revolution spawned numerous technological advances that were crucial in bringing new land into production. For, contrary to conditions in Europe, in America land was plentiful but hands to work it were scarce.

Always quick to adopt labor-saving machinery, Thomas Jefferson himself designed an award-winning "mouldboard of least resistance." Later inventors designed special plows for special tasks — breaking ground, turning furrows on a slope, preparing seedbeds, burying stubble. Reduced draft led to the sulky plow, which a farmer could ride. Drawn by four horses, a sulky gang plow pulling two bottoms doubled the acreage he could work while seated. Gang plows pulling a dozen bottoms were drawn by steam tractors. Planting and weeding were facilitated by spring-tooth harrows, end-gate seeders, grain drills, and straddle-row cultivators.

Still, a farmer cannot profit by planting more than he can harvest in good season. The harvesting of row crops, such as tobacco, corn, cotton and vegetables, resisted mechanization, but nineteenth century inventions revolutionized the harvesting of small grain. Around 1800 the sickle began to be replaced by the more efficient cradle, a scythe with a frame to catch the cut grain. Cradle swingers, cutting two acres a day, would bring in most of the grain harvest for the next fifty years.

In the 1830s Cyrus McCormick invented a mechanical reaper that set the pattern for later models, and Obed Hussey designed a cutter bar that is still used today. Early reapers could cut 15 acres a day, but farmers wanted machines to take on additional tasks. A self-raking reaper swept measured amounts of grain off the platform for field hands to bind into sheaves. "It was hard to believe that anything more cunning would ever come to claim the farmer's money," Garland reminisced, looking back on the harvest of 1874. But self-binding reapers and more were already cutting a swath across the midwestern prairie.

As early as 1836, Hiram Moore and J. Hascall had built a harvester that not only cut and threshed grain in the field but cleaned and bagged it as well. Powered by ground wheels, it could harvest 25 acres a day. By the 1880s large-scale combines, equipped with 18-foot cutters and drawn by teams of twenty mules or horses, could harvest as many as 45 acres a day. Combines armed with 30-foot cutters and pulled by steam tractors were capable of cutting 90 acres a day, but steam engines offered little economy of labor because so many hands were needed to fire them. The gasoline engine would spark the next great technological explosion.

The first internal combustion tractor to prove useful in the field was a jerry-built contraption that completed a 50-day threshing run in the harvest of 1892. Standardization and improved design led to mass production, and by 1914 more than 17,000 gasoline tractors were in use on American farms. Small, powerful, and easy to maneuver, a gasoline tractor could be harnessed to perform countless chores besides plowing and harvesting. A farmer could use the tractor belt to run a water pump, a chain saw, a grinding mill, even his wife's washing machine. A power take-off attachment would enable him to operate a baler, a sprayer, or other mobile equipment. A gasoline tractor worked tirelessly from sunup to sundown, and did not have to be watered, fed, and bedded when day was done. The labor saved from caring for draft animals could be invested in tilling the soil, and the land freed from growing fodder — about a quarter of our cultivated land was devoted to feeding mules and horses when their numbers peaked around 1920 — could be shifted to growing food crops.

Large-scale mechanization triggered a major agrarian assault on the Great Plains. Mixed- and short-grass prairie had once supported as many as 30 to 70 million bison, or American buffalo, in this semiarid region. Seeking to exploit the native vegetation, cattlemen had flooded the unfenced plains with Texas longhorns in the 1870s. But overgrazing, collapse of the market, drought, killer blizzards, and, not least, the advent of barbed wire, conspired to bring an end to the open range in the 1880s. Buoyed by periods of above-average rainfall, successive waves of homesteaders sought to convert the prairie to cropland only to be defeated by recurring drought. Railroad hucksters and dry-farming enthusiasts inspired a surge of settlement around 1900. Optimists held that "rainfall follows the plow" — or the railroad or telegraph lines, or even military battles. C.W. Post, better known as a captain of the breakfast food industry, mounted a rain-making offensive in which he detonated great quantities of dynamite in a series of mock engagements. But chronic drought persisted, and dust storms followed.

Undeterred, farmers continued to move to the high dry plains, especially after Congress threw open vast tracts of western land for 640-acre homesteads. During World War I demand for farm products rose sharply, and between 1914 and 1919 land in wheat grew by 50 percent, as prices soared to record levels. Riding out intermittent farm depression, wheat expansion gathered momentum in the 1920s. Tractors pulling a dozen disk plows over the broad flat plains helped reduce the region's production costs to a nationwide low. Good weather held, and bumper crops brought big profits.

Then came the "Dirty Thirties," when those who had sown the wind reaped the whirlwind. First the strike, a global economic depression eroded the price of wheat by two-thirds in two years. Desperate for cash to cover operating costs and debt payments, overextended plains farmers converted still more acreage to wheat. Then in 1931 a severe and protracted drought set in, with catastrophic consequences.

Stripped of the year-round cover of native grasses and unprotected by withered wheat and cotton, the parched prairie soils lay exposed to winds that swept unimpeded across the treeless plains. The farmer's prime resource blew away in clouds that could carry off hundreds of millions of tons of fertile topsoil in a single storm. "Black blizzards" reduced visibility to zero and deposited dust in drifts up to 25 feet high, burying stunted crops, idled tractors, empty barns, and abandoned houses. Hardest hit was a 25,000-square-mile area of the southern plains that became known as the Dust Bowl, but wind erosion scarred every region of the nation's heartland.

Communities collapsed as banks foreclosed, and farmers joined the ranks of farm laborers and tenants who had been "tractored out," displaced by "snub-nosed monsters, raising the dust and sticking their snouts into it, straight down the country," as John Steinbeck portrayed the new technology in *The Grapes of Wrath*. In this moving epic of Oklahoma's dispossessed, he evoked the human losses of the Dust Bowl

Leaning into the wind, a homesteader and his sons head for the flimsy shelter of a root cellar in
Cimarron County, Oklahoma, in the heart of the Dust Bowl. With no trace of a crop and the landscape
smothered in drifts, earth and sky become one as great clouds of dust swirl up in this 1936 storm.

tragedy: "Behind the tractor rolled the shining disks, cutting the earth with blades – not plowing but surgery . . . And pulled behind the disks, the harrows combing with iron teeth . . . Behind the harrows, the long seeders – twelve curved iron penes erected in the foundry, orgasms set by gears, raping methodically, raping without passion . . . And when that crop grew, and was harvested, no man had crumbled a hot clod in his fingers and let the earth sift past his fingertips. No man had touched the seed, or lusted for the growth."

Scarcely a century after the New World was first put to the plow, voices had warned of the need to conserve the soil and its fertility. In 1768 George Washington, then a concerned Virginia farmer, criticized the practice of continuously cropping corn and wheat "until the land is exhausted, when it is turned out, without being sown with grass seeds, or any other method taken to restore it . . . Our lands were originally very good; but use, and abuse, have made them quite otherwise." Another Virginia farmer, Thomas Jefferson, recommended plowing on the contour and planting clover to retain fertile soil. But such voices were rare, and few listened.

In 1929 alarms sounded by soil scientist Hugh Hammond Bennett aroused Congress to fund a soil erosion survey. But not until a dust storm reached continental proportions in May of 1934 did the American people awaken to the peril. "This particular dust storm," Bennett recalled, "blotted out the sun over the nation's capital, drove grit between the teeth of New Yorkers, and scattered dust on the decks of ships 200 miles out to sea. I suspect that when people along the seaboard of the eastern United States began to taste fresh soil from the plains 2,000 miles away, many of them realized for the first time that somewhere something had gone wrong with the land."

The Great Plains was not the only place something had gone wrong, and wind erosion was not the only culprit. Across the land, water erosion was rampant, scouring off sheets of unprotected topsoil, etching rills in denuded slopes, gouging out gullies in abused farmland. In 1939 Bennett estimated that erosion was destructively active on some 775,000,000 acres – more than a third of the nation's total land area.

Erosion is a natural phenomenon caused by wind and running water, but it can be severely aggravated by human activity. In pursuit of abundance, farmers had plowed their way across the continent, too often careless of long-term effects on the land because of an enduring belief that the land was limitless. The environ-

Farmers take a midday break on threshing day to savor the good earth's bounty, in a scene that Grant Wood painted in 1934 but that evokes his Iowa boyhood at the turn of the century. In late summer, after the grain had been cut and stacked in shocks, the itinerant thresher would arrive and, with a shriek of its steam whistle, summon neighboring farmers to help in the work. At the stroke of noon on threshing day, hungry crews came in from the fields to a feast the farmwife and her daughters had spent days preparing. Tables were placed end to end and every chair, including the piano stool, was pressed into service. Here, the hired hand and the family's sons, washing up on the porch, will have to wait for a second sitting. Help at harvest time was repaid in kind as the custom thresher moved from farm to farm. As in barn raisings and quilting bees, cooperation was a tradition of American farm life.

mental crisis that confronted American agriculture in the 1930s forced us to recognize that the land is finite, that it is fragile and irreplaceable. All that sustains us springs from the soil. Without it, there can be no agriculture.

We also realized that, just as we had created the crisis by abusing the land, it was within our power to control erosion by caring for the land. In 1935 Congress established the Soil Conservation Service (SCS) as a permanent agency. Providing both technical and financial assistance, the SCS helped farmers take appropriate steps to reduce soil erosion. Effective techniques included alternating crops in strips and establishing shelterbelts to create wind barriers, plowing on the contour and constructing waterways to control runoff, choosing suitable crops for terrain and climate, and practicing conservation tillage.

For three centuries American agriculture had increased production by increasing labor productivity and bringing more and more land under cultivation. Then the frontier closed, and as attitudes towards the land began to shift from exploitation to conservation, the focus narrowed to productivity per acre. The spiraling food needs of World War II and the postwar years were matched by improved seed and stock, greater farm mechanization, expanding irrigation, and increased use of fertilizers, herbicides and pesticides. As a result, in the half century from 1930 to 1980 farmers more than doubled their output while cropland acreage remained essentially unchanged.

But there can be too much of a good thing. Chronic crop surpluses depress prices rather than raise profits. Increasingly sophisticated equipment causes costs to skyrocket. Escalating chemical inputs bring diminishing returns. Improvident irrigation not only depletes water resources but can ruin farmland. Even high commodity prices have their downside: They provide an incentive to cultivate vulnerable land, with the result that soil erosion has repeatedly risen to levels that approach or exceed those of the 1930s.

The last quarter of the twentieth century has brought increasing awareness of threats to the health and continued growth of American agriculture. As the world's most efficient producer of food, grow it must, just as the world's need for food continues to grow; in the year 2000 there will be half again as many mouths to

feed as there were in 1976. Consciousness of environmental constraints, however, has added a new dimension, time, to measures of agricultural productivity. The challenge for American farmers today is not to achieve ever greater abundance but to assure abundance for generations to come. They are meeting this new challenge, as they have in the past, by seeking new solutions, with the help of chemists and physiologists, geneticists and biotechnologists, policymakers and planners. At the urging of environmentalists, ecologists, and latter-day prophets, they are also rediscovering old ways of farming born of experience and their deep and abiding love of the land.

At the beginning of the century Liberty Hyde Bailey, botanist, educator, and farmboy-turned-philosopher who saw in nature "a constant, intricate, friendly beauty," pondered earth's beneficence and our indifference: "So bountiful hath been the earth and so securely have we drawn from it our substance, that we have taken it all for granted . . . with little care or conscious thought of the consequences of our use of it; nor have we very much considered the essential relation that we bear to it as living parts in the vast creation." Prodigal children of Mother Earth, at the end of the century we are beginning to make amends and seek a state of harmony in a world of finite resources, beginning to realize we must reciprocate earth's generosity and nurture the beautiful, bountiful land that nourishes us.

THE STAFFS OF LIFE

In a tribute to corn in 1864, the *Prairie Farmer* declared multifaceted maize to be "unequalled for fattening purposes. The buxom girls and the stalwart sons of the West deem it the staff of life when made into bread, and when made into whisky many think it life itself." Fried, boiled, baked, roasted, corn was eaten by frontier families on and off the cob, and as mush, pudding, corn bread, grits, johnnycake, spoon bread, hoecakes, corn pone... Distilled, corn mash became corn likker, a deep swig of which could "stop the victim's watch, snap his suspenders and crack his glass eye right across," according to raconteur Irvin S. Cobb. Boys smoked corn silk behind the barn; girls cradled corn-husk dolls. Cornstalks fattened the hogs. And dried corncobs, rough but absorbent, stocked American outhouses into this century. Today modern milling transforms corn into oil, margarine, sweeteners, starch, and an unsuspected ingredient in numberless products: paint, paper, plastics, tires, safety glass, fuel, cosmetics, peanut butter, instant coffee...

Corn, wheat, and rice are the world's most important food crops. These and other grain-bearing grasses – barley, oats, sorghum, millet – supply more than half of mankind's caloric intake, or as much as three-quarters counting the meat, milk, and eggs derived from grain-fed livestock. Another major staple is the soybean, a legume that is a key source of protein in Asian diets and animal rations. These four staffs of life, together with three root staples – potatoes, sweet potatoes, and cassava – are the crops that enable the world's burgeoning population to stave off starvation.

About a quarter of America's cropland is devoted to corn, a fifth of it to wheat, another fifth to soybeans, and a small but significant fraction to rice. Cultivation of these crops follows a regional pattern. The Corn Belt stretches from Ohio into Nebraska, where fertile soils and ample rainfall meet corn's requirements; soybeans serve as a rotation crop. Wheat, less demanding, dominates peripheral areas marked by lighter soils and less precipitation. On the Southern Plains, winter wheat is sown in the fall and harvested in early summer; on the Northern Plains, spring wheat reigns. In the semiarid West, wheat is planted only in alternate years and the land left fallow between crops to accumulate moisture. Soybeans are grown throughout the South, on more acres than cotton when cotton was king. Rice is cultivated in regions with abundant water and a long, hot growing season: the Delta states, the Gulf coast of Texas, and California.

The United States grows huge quantities of these staples – far more than we consume. We export more than 20 percent of our corn and between 50 and 60 percent of our soybeans, wheat, and rice. More than 100 countries depend on our surpluses for human or animal nutrition, and dependence is increasing – America has become the world's breadbasket and feed trough. Keeping bins filled, per-acre yield has more than doubled for wheat and more than quadrupled for corn since 1940, thanks to expanded irrigation, greater use of fertilizer, better farming practices, and improved seed.

Thomas Jefferson held that "the greatest service which can be rendered any country is to add a useful plant to its culture, especially a bread grain." Native Americans added corn to the world's grain inventory, and others since have served mankind by developing high-yielding varieties of staple crops. But reliance on so few cultivars, bountiful as they are, carries risks; their genetic uniformity has made them vulnerable – to epidemics, insect plagues, environmental stress. Plant scientists can breed new varieties, they can transplant and even alter genes, but their raw material must come from existing germ plasm. Many earlier strains and wild ancestors of the world's major food crops are threatened with extinction. To insure the future of the staffs of life, we must preserve the diversity of earth's genetic resources.

Left: Gently swelling, verdant and fertile, no place on earth is so well endowed agriculturally as America's Corn Belt. On a dairy farm in north central Ohio, dense stands of corn alternate with strips of freshly mown hay, and beyond, bands of wheat just beginning to ripen are tinged with gold. Per-acre yields of grain grown for food and feed in Ohio are among the highest in the nation.

● *Right*

GOLDEN WHEATFIELDS, silver threads of pea-vine residue, and velvety black soils coalesce to form a star in the Palouse River basin, near Colfax, Washington. A product of wind erosion, the deep loess soils are highly susceptible to water erosion when lying fallow. To protect them, the farmer, instead of using a moldboard plow to turn the pea stubble under, is using a chisel plow to loosen the subsoil and only lightly stir in the crop residues as stubble mulch. In zones of relatively low precipitation, mulching is a highly effective method both of controlling erosion and of conserving moisture.

● *Right*

CONVOLUTED STRIPS of beige winter wheat, green-gold spring barley, and dark summer fallow are drained by grassed waterways near Cottonwood, Idaho. To control erosion of the rich silt loam, the Tacke family has maintained this contour stripcropping system for more than a quarter of a century, although it complicates farming operations immensely.

● *Previous pages*

CONVERGING BUT never merging, strips of wheat and fallow near Great Falls, Montana, are cropped in alternate years, using two years' precipitation to produce one harvest. To reduce erosion, the strips are laid out counter to prevailing winds, as is the stand of trees that shelters the farmstead.

● *Left*

LIKE A shawl cast casually on uneven ground, strips of partly harvested corn and alfalfa hay cling to the land's contours in northwestern Illinois. The contour strips help control erosion, as does crop rotation, the most common sequence in the area being two years of corn, one year of oats, and three years of sod-forming forage. Seeking to profit from this crop mix, many farmers use the corn as feed for dairy operations or for raising hogs or beef cattle.

● *Left*

SEPTEMBER COLORS the knolls and folds of Berks County, Pennsylvania, with the ruddy hue of ripening corn, the soft browns and beiges of small grain stubble, the bright green of hay crops, and the deep green of dense woods. Cultivating along the contours and alternating strips of row crops, small grain, and forage helps prevent erosion by slowing the flow of runoff down the slopes. The narrow green bands are grassy swales, or shallow channels, that intercept runoff and divert it to a safe outlet. Crop rotation in Berks County is generally two to three years of corn, soybeans, wheat, barley, or oats, and four to five years of alfalfa hay. Some of the small grain may be processed as food, but most of the crops will be used as livestock feed on the farm or sold on the open market for commerical feed.

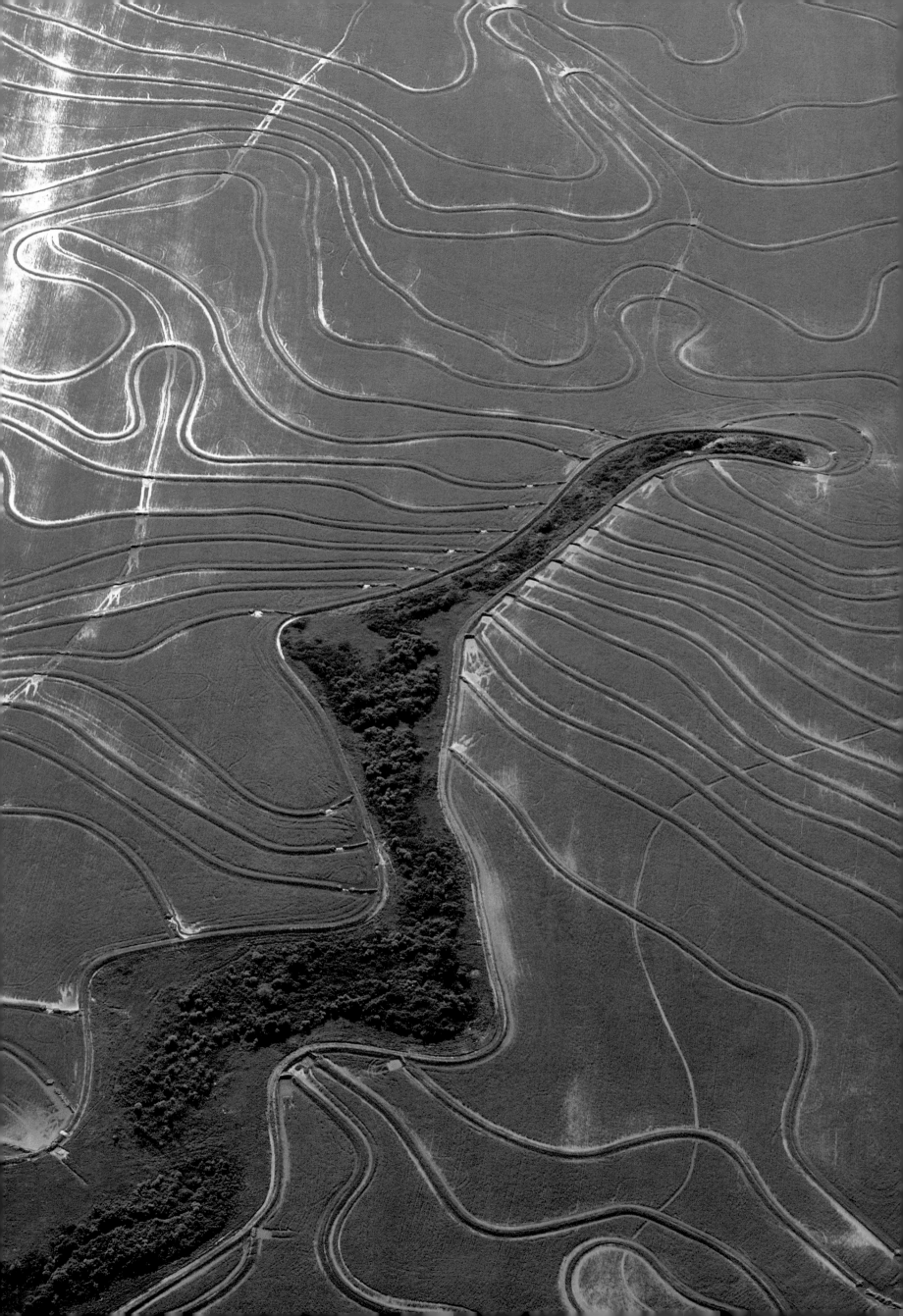

● *Left*

CONTOUR LEVEES keep rice fields
flooded to a depth of two to four
inches during most of the 130-day
growing season in the Gulf Coast
region of Texas. A field-border
levee prevents the water from
running off into a gully that snakes
down the slope, and levee gates
control the flow between fields.
The Texas rice industry uses as
much as 20 percent of the state's
renewable water resouces, but in
the face of increasing urban and
industrial competition, growers are
cutting back on water use and
getting higher yields by using
shallower flood depths to grow
new early maturing semidwarf
rice varieties.

● *Above*

MYRIAD MEANDERING levees echo the
irregular topography of rice fields
near Stuttgart, Arkansas, but the
difference in elevation between
the contour levees is only about 0.2
feet. When the rice begins to tiller,
or put forth shoots, the fields will
be flooded and remain so until just
before harvest. Thanks to an
efficient air passage from shoot to
root, rice thrives in a water-logged
environment. Much of the acreage
devoted to rice in the South has
been reclaimed from wetlands.
Deep and slowly permeable soils,
gently sloping topography, and
access to ample water make
eastern Arkansas prime rice-
growing country.

● *Overleaf*

SUNRISE GILDS waves of corn carved
out by a farmer when he "opened"
his fields with a combine to start
the harvest in southwestern Iowa.
The pattern follows the contours of
level terraces built to control
erosion of the deep loess soil,
which helps make Iowa almost
bewilderingly bountiful, with
average yields of about 130 bushels
of corn per acre; many farmers
regularly produce more than 200
bushels per acre and some have
topped 300 bushels per acre of this
native American grain. Prized as
"the prince of grasses" by
agricultural geneticist Paul
Mangelsdorf, corn has also been
called "the king of all farm crops."

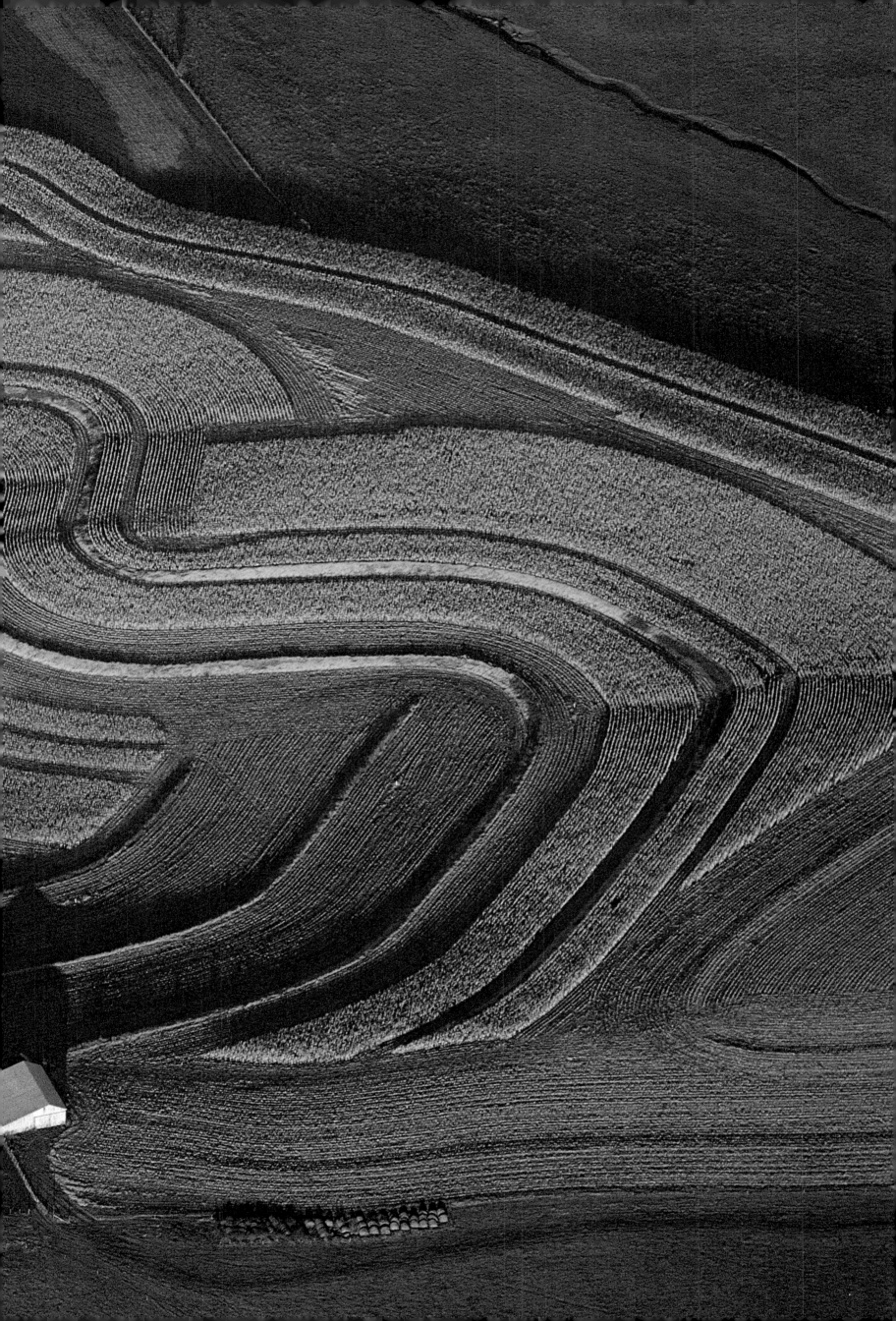

● *Above*

RIPE RICE ripples across close-set levees on a stretch of steep terrain in Arkansas, where one year of rice is usually rotated with two years of soybeans. The upper field has already been harvested and prepared for replanting.

● *Right*

LODGED RICE makes the going rough for harvesting combines near Greenville, Mississippi. Lodging often occurs after fields are drained for harvest, when hollow rice stems cannot support heads heavy with grain in wind or rainstorms.

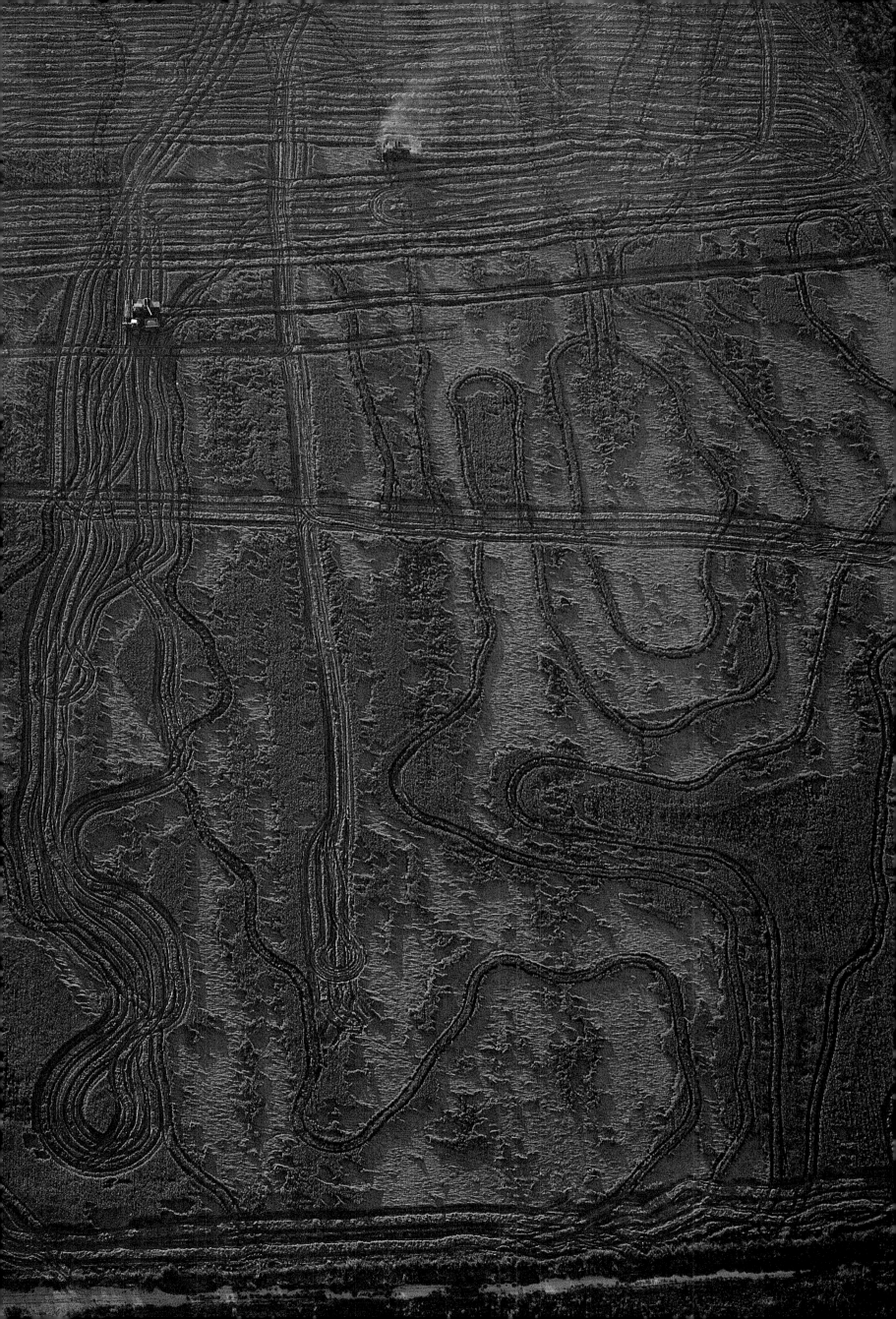

● *Right*

IMPRINTED ON the land, the farmer's hand calms the "stormy sea" of southeastern Washington's Palouse with alternate strips of pale gold wheat, darker barley, and gray-brown fallow. Contour stripcropping checks erosion on the turbulent terrain by intercepting runoff, as do the stepped level terraces, or dikes, at left. The region's fertile soils evolved from volcanic deposits and wind-blown loess.. Beginning about 26 million years ago, huge lava flows hardened into thick basalt covering parts of Washington, Idaho, and Oregon. Later, as the basalt weathered, winds from the southwest swept up the fine, loose soil and deposited it in the dunelike hills that now dominate the landscape.

● *Right*

AS THE golden brown harvest is gathered in from the hills and hollows of Palouse country, straw and stubble are left behind to protect the vulnerable soil through the winter and spring; land now fallow will be planted for next year's crop. The region was christened by French explorers, who likened the native grass carpeting the land to an Old World lawn, or *pelouse,* but settlers soon transformed the Palouse prairie into a granary. Because the Cascade Mountains cast a rain shadow across eastern Washington, farmers have developed a pattern of dry farming keyed mainly to wheat, which has a low demand for water compared to other major grains. Palouse growers usually use a two-year cropping cycle, alternating winter wheat with fallow, which stores up moisture for a season, or sometimes a three-year rotation of winter wheat, spring barley, and fallow.

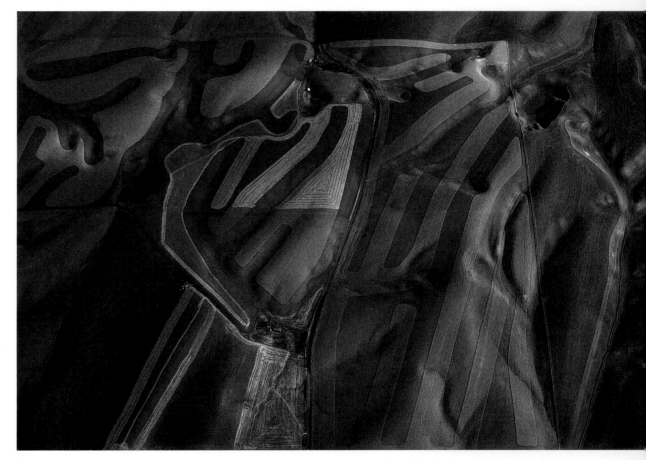

● Left

HILLTOPS PLANTED in winter wheat are banded by contour strips of green lentils and brown barley to control erosion near Pullman, Washington. Despite low precipitation in the Palouse, the steep slopes combined with highly erodible loess soils can produce soil erosion rates as high as 200 tons per acre per year if the land is farmed up- and downhill. Sudden snowmelt or spring rain on saturated ground can carry off whole sections of topsoil in mini landslides. Recent losses of 1,200 pounds of topsoil for every bushel of wheat harvested make soil-conserving practices such as contour stripcropping essential if farming is to continue to flourish in the region.

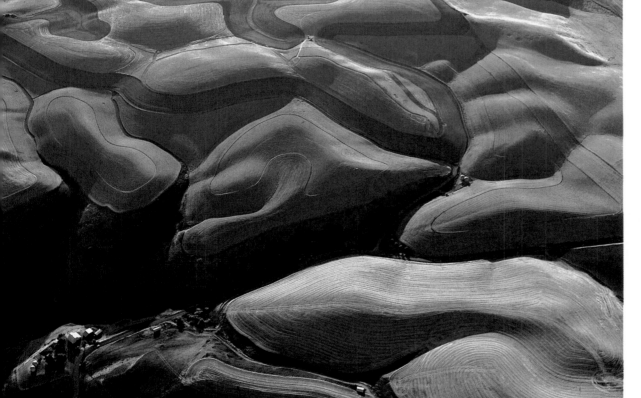

● Left

WHALEBACK HILLS of the Palouse are mantled with spring barley on the crests and winter wheat along the flanks, with one leviathan already harvested. Washington farmers harvest about two million acres of wheat annually, with a yield of about 60 bushels per acre, almost double the national average, the bulk of it white wheat, which is used for cakes, cookies, pastries, crackers, and noodles.

● Overleaf

COLLAGE OF warm earth tones is made up of soil- and water-conserving strips of ripe grain, sun-bleached stubble, and dark fallow near Pomeroy, Washington. This distinctively American panorama is the work of a careful farmer meeting the needs both of his crops and of the land, but only one in ten Palouse farmers has adopted such practices because of time, expense, and difficulty in using wide-swath equipment.

ANIMAL HUSBANDRY

The longhorn steer reigned as king of the range for only two decades, but the cowboy lives on in legend and song as a bronzed, bow-legged knight-errant, "a man with guts 'n' a hoss" driving herds of bellowing cattle from Texas north to the Great Plains, to be shipped to eastern markets and to stock western grasslands. As the prairie began to green in 1866, the first Long Drive of 260,000 longhorns set out in bands of a thousand herded by half a dozen cowboys armed with lassos and six-shooters. Hazards of the drive included trails slippery with mud or choked with dust, rain-swollen rivers, dry waterholes. Hostile Indians stampeded herds; rifle-toting settlers erected barricades. Relatively few of those trail-blazing longhorns reached the railhead in Missouri, but those that did fetched ten times over what they had cost in Texas – and the greed stampede began.

In scarcely a score of years the cowboy's realm spread throughout the West as speculators chased profits. Then even more suddenly the empire collapsed, from overconfidence, overstocking, overgrazing. The winter of 1886–87 dealt a mortal blow: Deep snows, howling blizzards, and temperatures as low as 68 degrees below zero left the plains littered with carcasses and starving survivors. In the wake of this devastation, cattle barons of the open range became ranchers, husbanding smaller herds of carefully bred stock in fenced pastures. And foot-loose cowboys became hired hands, pitching hay, shoveling manure . . .

Most of the beef Americans eat still comes from the Great Plains. But today cattle are shipped there from many regions for fattening in huge commercial feedlots. With irrigation, the dry central plains provide abundant feed grain. Cattle are also finished in the Corn Belt, the region that produces most of the nation's hogs. In the mountain states, cattle and sheep migrate between high summer pasture and lower winter range. In Florida and California, mild winters and plentiful water make it possible to grow forage for beef and dairy cattle year-round. Dairy cows remain concentrated, however, in the Northeast and in farms around the Great Lakes.

Of the five billion broiler chickens produced in the United States each year, about three-quarters are grown in the South, where small farmers began contracting to raise chicks in the 1930s. Today the mom-and-pop broiler business has been rolled into a vast, vertically integrated industry, and poultry has stolen the lead from beef in the American diet. Another fast-growing southern specialty is aquaculture, notably catfish production in Delta flatlands, easily bulldozed into ponds.

Historically, livestock fit well in the economic pattern of the diversified family farm, feeding on crop residues and in turn providing power as well as meat, milk, eggs, and useful byproducts, including manure. Today livestock production has become highly specialized, and the cattle industry in particular draws heavily on agricultural resources. Pasture and range occupy more than a third of the nation's land area. In addition, more than half of the cropland is devoted to growing feed. And livestock account for more than half of our water consumption, mostly through irrigation of feed crops. To produce one pound of grain-fed beef, it takes from five to eight pounds of rations and some 2,500 gallons of water. Only two or three pounds of feed are needed for one pound of chicken, while catfish convert food pellets into fillets at a weight ratio approaching one to one.

Cattle's drain on cropland and water has focused attention on efforts to reinvigorate rangeland by reseeding, balancing stock to available forage, and adopting rotational grazing patterns. Our grasslands evolved in symbiosis with immense herds of buffalo, and perennial grasses thrive under periodic grazing; thanks to their extensive root systems, they can regenerate even when closely cropped. By going back to grass roots and carefully managing our resource endowment, we can raise ruminant stock in an ecologically sound and sustainable manner.

Left: Cattle fan out to forage on a snow-covered field in Minnesota, where hardy Herefords spend even the severe winters of the northern plains in the open. They feed on crop residues and on rations hauled out to them; in extreme cold, livestock burn so many calories that they require extra feed just to maintain body weight. Cattle wintering on the range in remote areas must sometimes be supplied by air drops after heavy snowfalls. Blizzards late in the season, after cattle have shed their winter coats, can spell ruin for stockmen.

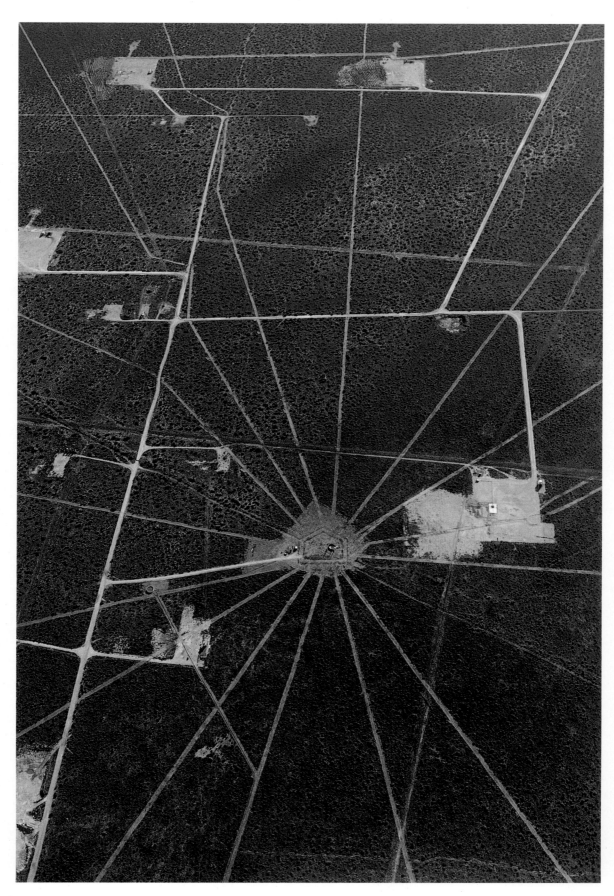

• *Above* _____ • *Right* _____ _____

SPOKES OF a cellular grazing system transect the grid of an oilfield near Wink, Texas. Sectors of fenced pasture radiate from a central reservoir. Rotating cattle from one cell to the next every three days allows grasses to recover.

CROSSBRED CATTLE grow fat and lazy in a farm feedlot near Spink, South Dakota. Bought as 500-pound calves or 700-pound yearlings, they are fed concentrated rations to bring them to 1,100-1,200 pounds before they are sent to slaughter.

● *Right*

RIPPLING WINDROWS of hay are ringed by myrtle and cabbage palms, while bales spatter the field beyond, near Lake Okeechobee in Florida, where high forage production supports numerous cattle ranches. Most are "cow-calf" operations with breeding herds that produce a crop of calves each spring. In the fall the calves are often sold as stockers to western farmers and ranchers who feed them silage, hay, or crop residues or keep them on pasture over the winter. At fifteen months the animals are sold to feedlots on the Great Plains, where they are finished on a diet of grain for proper marbling.

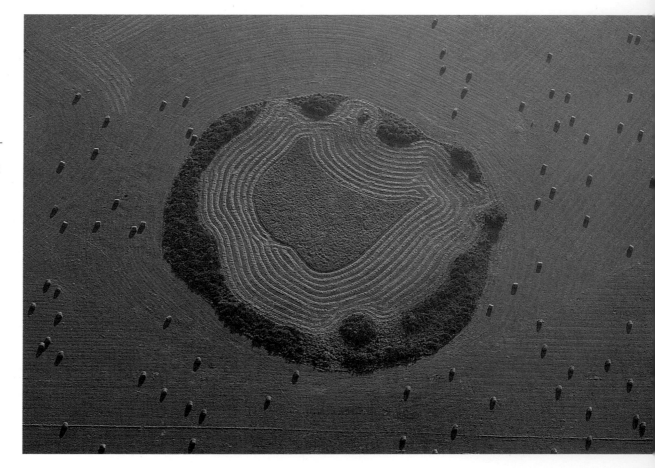

● *Right*

HOLSTEIN COWS feed on quality rations at a high-tech dairy farm in California's San Joaquin Valley. Keeping them on the job, stanchions lock automatically when cows put their heads through the feeding fence. Protein-rich alfalfa hay is central to their diet. Around 1900 a cow's annual output was about 3,000 pounds of milk; today cows in California top 18,000 pounds on average. In automated milking barns, their udders are plugged into pneumatic machines – "those strange tentacled calves with their rubber, glass, and metal lips," wrote William H. Gass – which suck them dry two or three times daily. A single technician can milk as many as eighty cows an hour. Neckchain transponders report yields to a computer that handles data for herd management.

● *Left*

PLASTIC BAGS filled with fermenting silage near Las Cruces, New Mexico, resemble bloated tube worms. Corn or alfalfa cut and bagged during the summer will provide succulent year-round feed for milk cows. Stacks of dry alfalfa hay represent another way of storing pasture for later consumption. Discovered in the 1870s, silage was widely used by American dairy farmers by 1900. Harvested green, a variety of crops – corn, grass, legumes, sorghum, small grains – can be preserved in a moist state by partial fermentation in a silo.

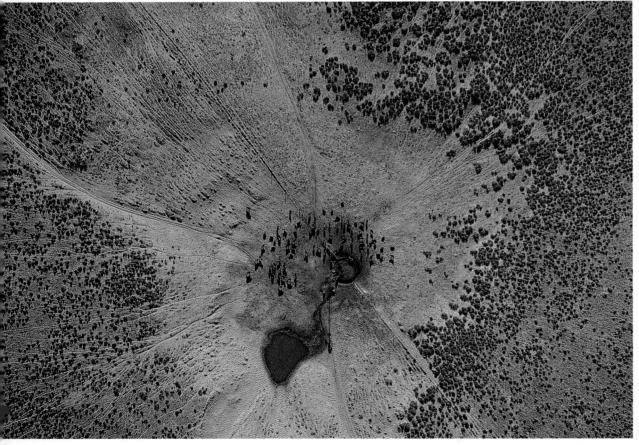

● *Left*

CATTLE CONGREGATE at a water tank, overgrazing and trampling semiarid rangeland near Greeley, Colorado. They feed on sagebrush and native grasses, which do not fall down and rot as in humid areas, but cure as "standing hay" for winter forage. Their multichambered stomachs enable cattle to harvest plants not digestible by humans and make economic use of land not suitable for cropping.

● *Overleaf*

SPRINKLERS SETTLE the dust at a feedlot with a capacity of 100,000 beef cattle at Greeley, Colorado. The animals are fed rations of fortified grain for about 120 days to produce the fatty tissue needed for tender cuts of meat that command premium prices. Most of the nation's large custom feedlots are concentrated on the central and southern Plains, close to abundant feed crops grown on land irrigated by the Ogallala Aquifer.

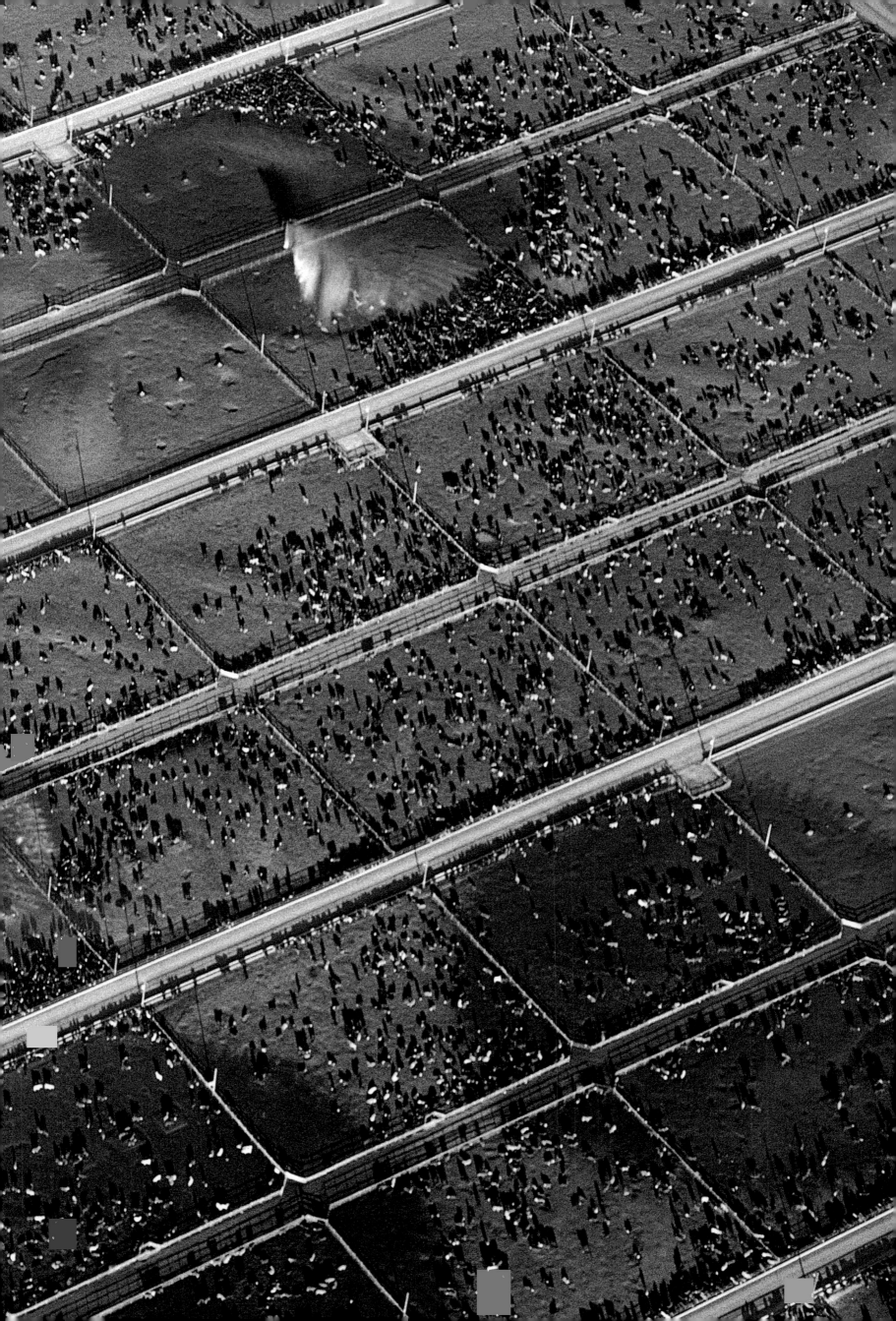

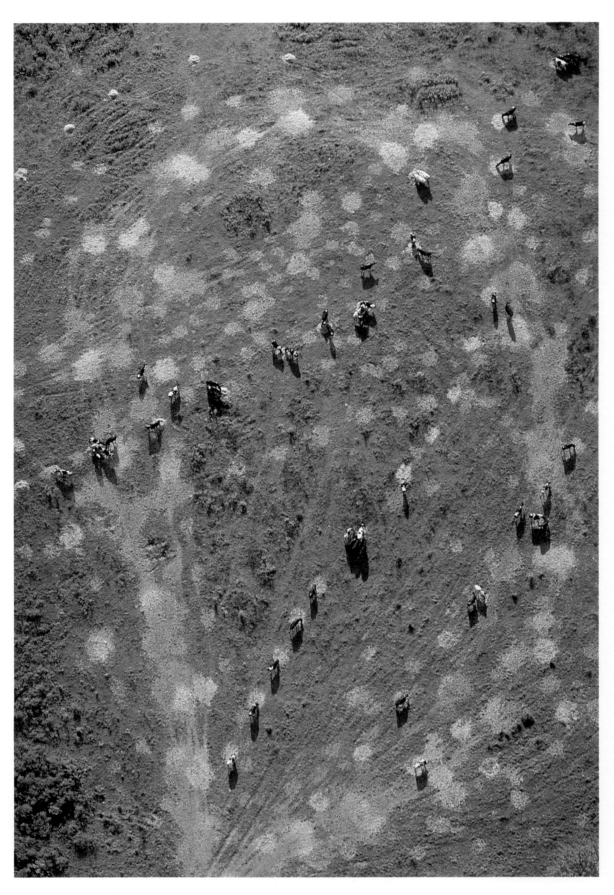

● *Above*

DISDAINING GREEN forage, dairy cows put out to pasture in upstate New York seek out piles of succulent corn silage. Supplemental feed boosts milk output, which averages 13,900 pounds per cow per year in New York.

● *Right*

SET LIKE an emerald solitaire amid maples and oaks, a fall alfalfa crop in the Hudson River valley will become high-protein winter fodder. Making hay while the sun shines, New York's 13,000 dairy farmers grow much of their feed.

62

● *Right*

LIKE FILINGS drawn to a magnet, turkeys flock around sheds and feeders on a Hutterite farm near Mitchell, South Dakota. Hutterische Brethren first broke the northern plains in 1874. Today, progressive Hutterite farmers, living in communes of 150 to 200 members, still seek new opportunities. Contracting to bring hatchlings to market weight, a community may raise 300,000 turkeys at a time. They may also grow grain for feed – corn and soybean meal with vitamin and mineral supplements – which can amount to 75 percent of production costs. Turkeys mature quickly. Most come to market at the tender age of 15 to 20 weeks, with a ready-to-cook weight of 12 to 20 pounds or more.

● *Right*

SPACED OUT for privacy, each farrowing house on a hog farm near Alcester, South Dakota, shelters a sow and her litter of pigs, and allows room to roam. Corn or oats growing in the surrounding field will supply feed grain and fodder. This type of husbandry requires more labor but less capital than factory farming, in which animals are confined year-round and the chores of feeding and cleaning largely automated. Selective breeding has redesigned yesterday's rotund hog into a longer, well-muscled animal with less than half as much lard, larger but leaner hams, and two more ribs, hence two more pork chops. It takes about 200 days to raise a piglet from birth to a 230-pound hog ready for market. The Midwest is home to most of the nation's 50 million hogs, animals that have paid off so many farms over the years they are fondly known as "mortgage lifters."

● *Left*

PAMPERED SHEEP graze on irrigated alfalfa, "Queen of Forages," in California's rain-short Coachella Valley. Sheeping down or hogging off – turning livestock loose on feed crops to save harvesting – is an old practice. Sheep, some say, improve later cuttings by eating more weeds than alfalfa. Come summer, this flock, like most of the nation's 10.8 million sheep and lambs, will be grazing on range-land. In the late 1800s, competition for grazing grounds on the Plains sometimes erupted in open war-fare between cowboys and sheep-herders, with well over half a million "woolies" as casualties – mostly driven over cliffs in Colorado-Wyoming cow country.

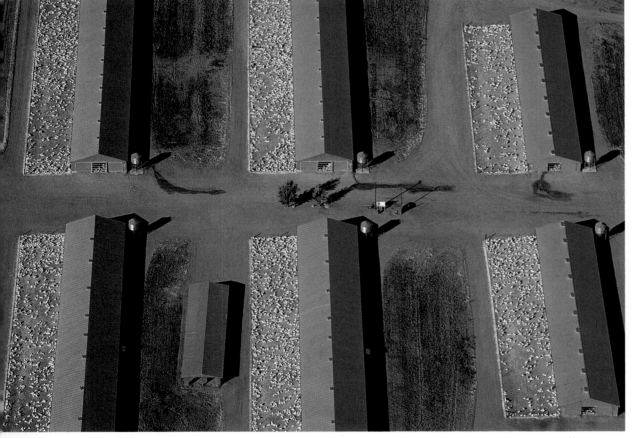

● *Left*

TURKEY BARNS with open runs near Greeley, Colorado, allow toms and hens to take the sun with little risk they'll fly the coop. Wild turkeys are strong fliers, but today's com-mercial turkeys are so large and heavy they can hardly walk, much less fly. Bred for plumper breasts, thicker thighs, and meatier drum-sticks, they cannot even mate naturally. They have also been bred for white feathers; colored feathers leave traces of pigment that make the skin look splotchy. Some 1,700 years ago, Indians of the Southwest bred turkeys for colorful plumage, which they used to make warm, lightweight robes. Today the U.S. produces almost a quarter of a billion turkeys annually. Thanks to aggressive marketing, low-calorie, low-cholesterol turkey now appears on menus year round. Still, about 45 million turkeys – one fifth of our production – are stuffed, trussed, and roasted, and consumed on Thanksgiving Day.

● *Above*

LIVESTOCK BARNS line Axberg Lake in
Minnesota. The three lower barns
produce 5,000 hogs annually. In
the upper barn, 120,000 pullets are
raised every twenty weeks to
replenish the farm's 320,000 hens.

● *Right*

RANCH HOUSES for breeding pairs of
silver foxes near Lehi, Utah, consist
of wooden dens that open on wire
pens roofed with sheet metal.
Foxes mate for life and each spring
produce four pups, on average.

66

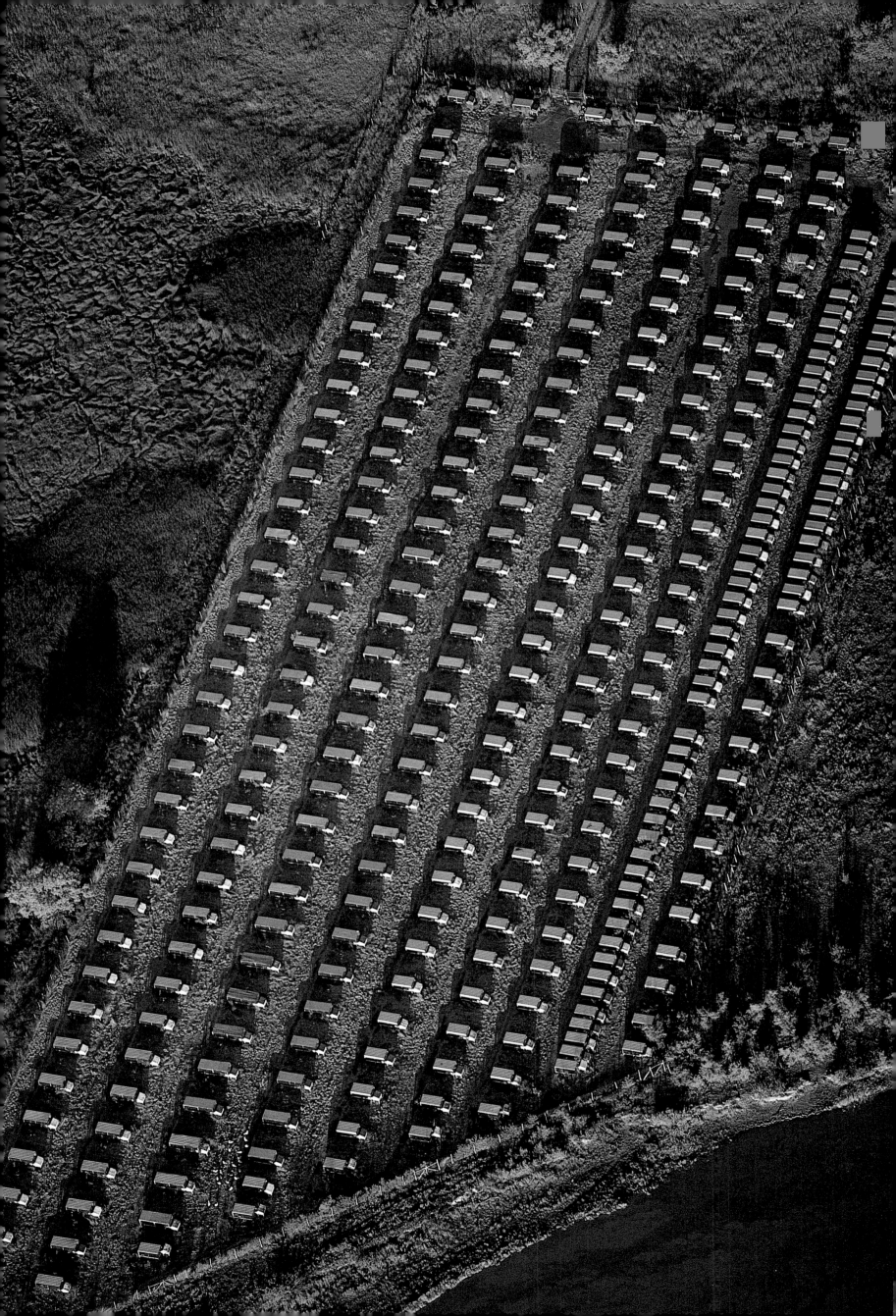

● *Above*

ERODIBLE KETTLE basins left by glaciers are spared on land cleared for forage production for dairying at Point MacKenzie, Alaska. Two bulldozers pulling an anchor chain knocked down dense stands of timber, then single dozers pushed the trees into berm rows, where they are repeatedly burned and eventually turned under. The fields will be planted to timothy, oats, and brome for silage, hay and pasture. On a typical Point MacKenzie dairy farm, a quarter of a million dollars is spent on land clearing – almost a fifth of the total investment, about the same as for livestock.

● *Facing page*

RIBBONS OF hay will be rolled into bales for winter feed for livestock in the eastern foothills of the Bighorn Mountains near Sheridan, Wyoming. The grass being harvested is probably a species of wheatgrass or brome, perennial drought-resistant bunchgrasses. Often used to reseed rangeland, they provide erosion control as well as choice forage. The presence of alkali salts, suggested by the white patches, would preclude cultivation of alfalfa. The field is bounded on one side by a ruler-straight property line, but the rest of its sinuous perimeter is determined by the rugged terrain.

● *Overleaf*

SINK SWAMP, near Fairfield, Utah, is an oasis of wet meadow rangeland and aquatic wildlife habitat in a semiarid region. Islands of vegetation are linked by trails used by duck hunters and cattle. Rushes and tufted hairgrass provide fair to good forage, but overgrazing can lead to the development of dense stands of less desirable sedges. A normal year yields about two tons of air-dried herbage per acre. Annual precipitation averages only eight to twelve inches, but the deep silt loam absorbs runoff slowly, and water collects in depressions, as in the sink at the left.

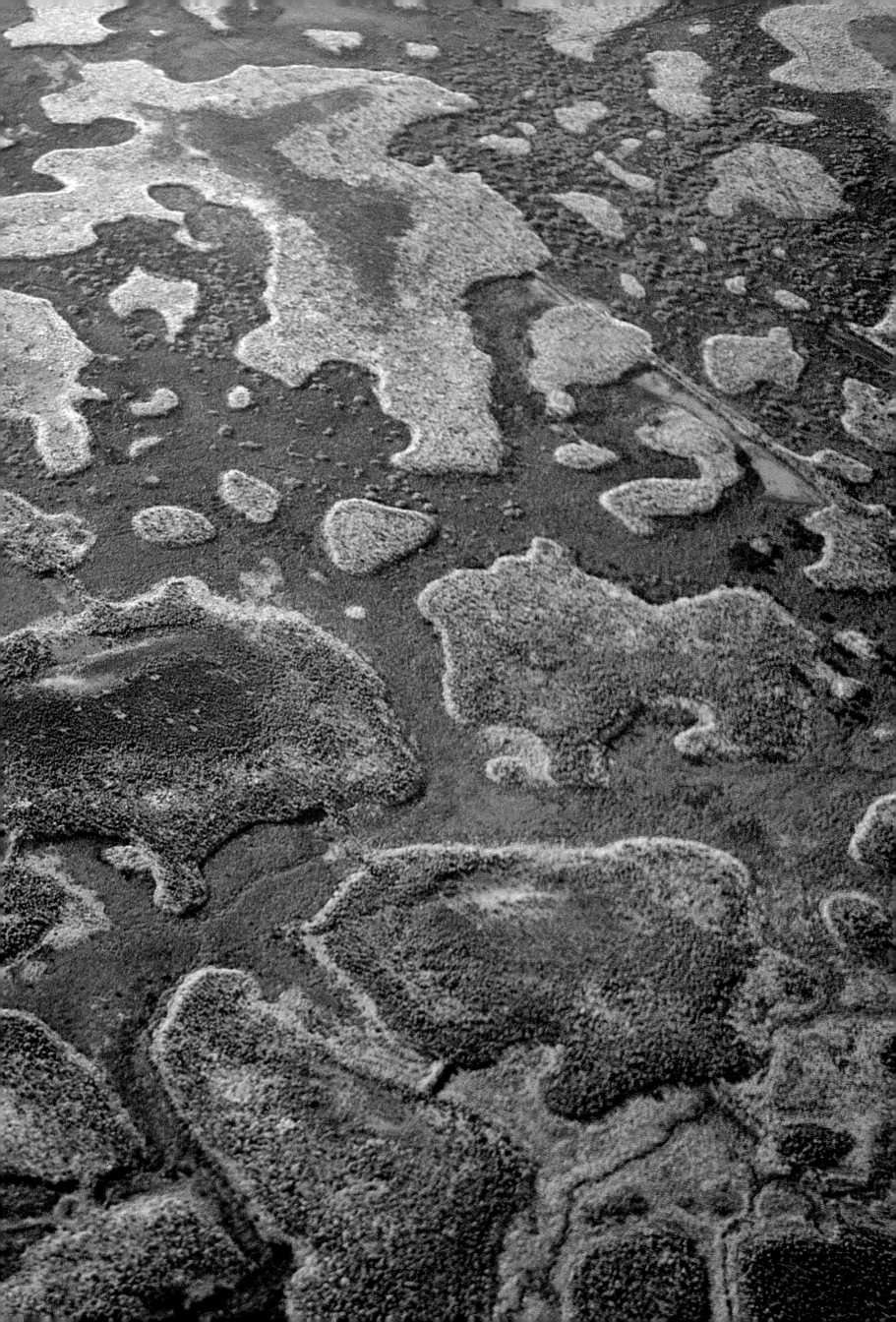

● *Above*

CULLED ORANGES and peels of
oranges squeezed for juice are
piled up by a tractor with a skip
loader after drying on an airfield in
California's Coachella Valley. Like
almond hulls, they are a low-cost
additive to livestock feed.

● *Right*

RED COUNTRY of Oklahoma shows
signs of abuse as former cropland
but is being improved as rangeland.
A diversion terrace now shunts
runoff to erosion control dams,
and smoothed-out gullies will be
seeded to native grasses.

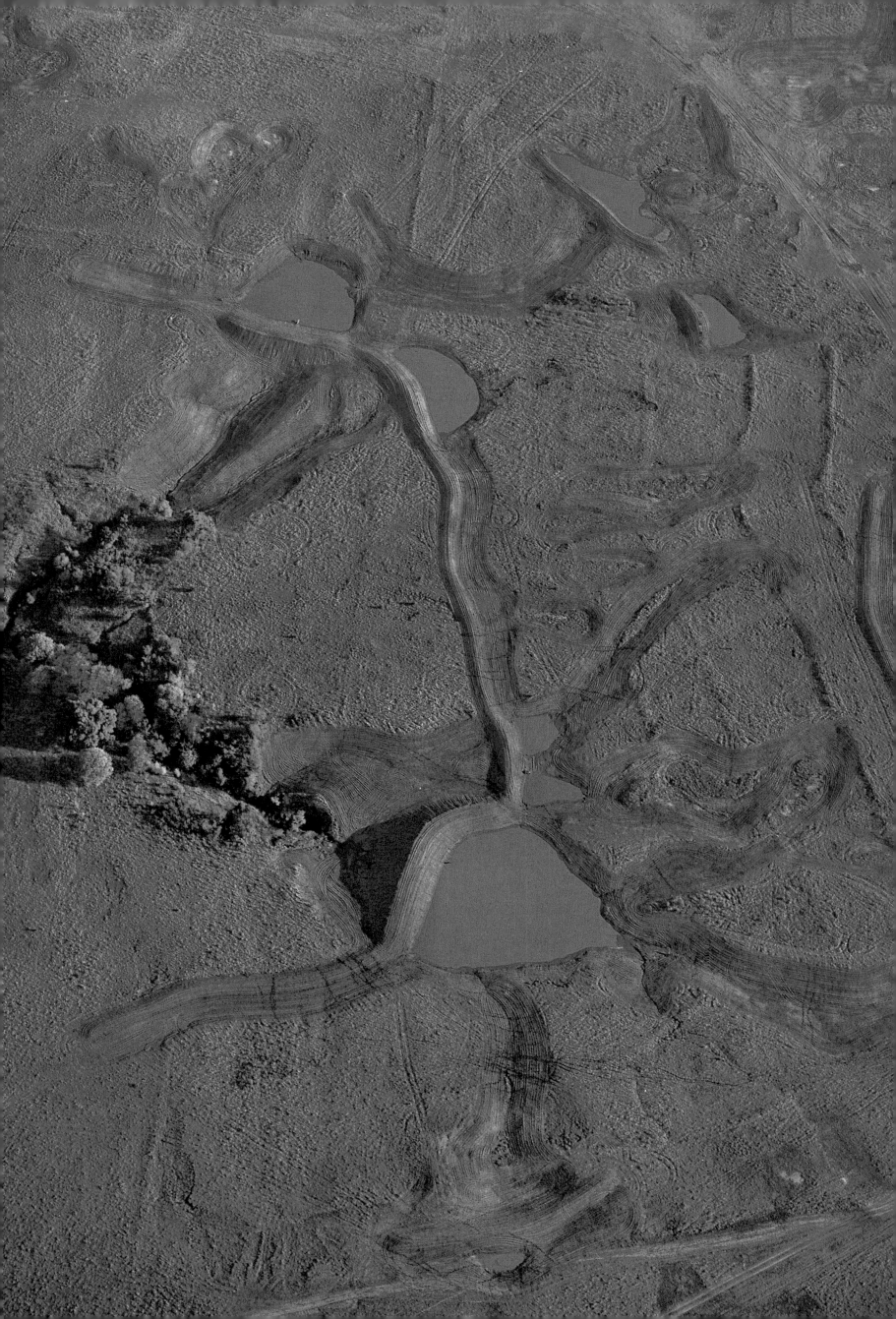

● *Above*

A PALETTE of greens marks aquaculture, which alternates with rice and soybean production in eastern Arkansas, an area with abundant spring water and clay flatlands to hold it. The dark pond probably brims with catfish, while the lighter ones swarm with baitfish, such as minnows or golden shiners. Yields of 5,000 pounds per acre can make catfish farming highly profitable, and production has increased more than tenfold over the last decade. But predators — racoons, snakes, but above all birds — are a growing problem; many, such as herons, are protected and can steal with impunity.

● *Facing page*

OLD RICE paddies serve as crawfish ponds and waterfowl preserves along the Black River in South Carolina. In the lower field, cross dikes circulate water for crawfish production. Ponds are sown with breeding stock and rice in the spring; young crustaceans hatch in the fall and grow fat on the rice crop. The harvest, from March to June, is labor-intensive, for crawfish must be trapped in baited pens, twenty-five or thirty to the acre, almost as their wild Cajun cousins are in the Louisiana bayous. Other aquaculture products include trout, salmon, hybrid striped bass, eels, oysters, mussels, shrimp, even caviar from sturgeon roe.

● *Overleaf*

AN OVAL training track dominates a horse farm in Kentucky bluegrass country, home of equine bluebloods. White board fences outline paddocks for exercising Thoroughbreds, the breed of champion racehorses. Today, horses are rarely used as draft animals in America, but the breeding of horses for sport and show, ranching and riding remains strong. Among an estimated 8.5 million head, the most popular breed is the all-purpose Quarter Horse, used for pleasure riding, cowherding, and racing at a quarter mile. Equestrian sports draw more than 110 million spectators annually, and betting on the ponies exceeds $13 billion.

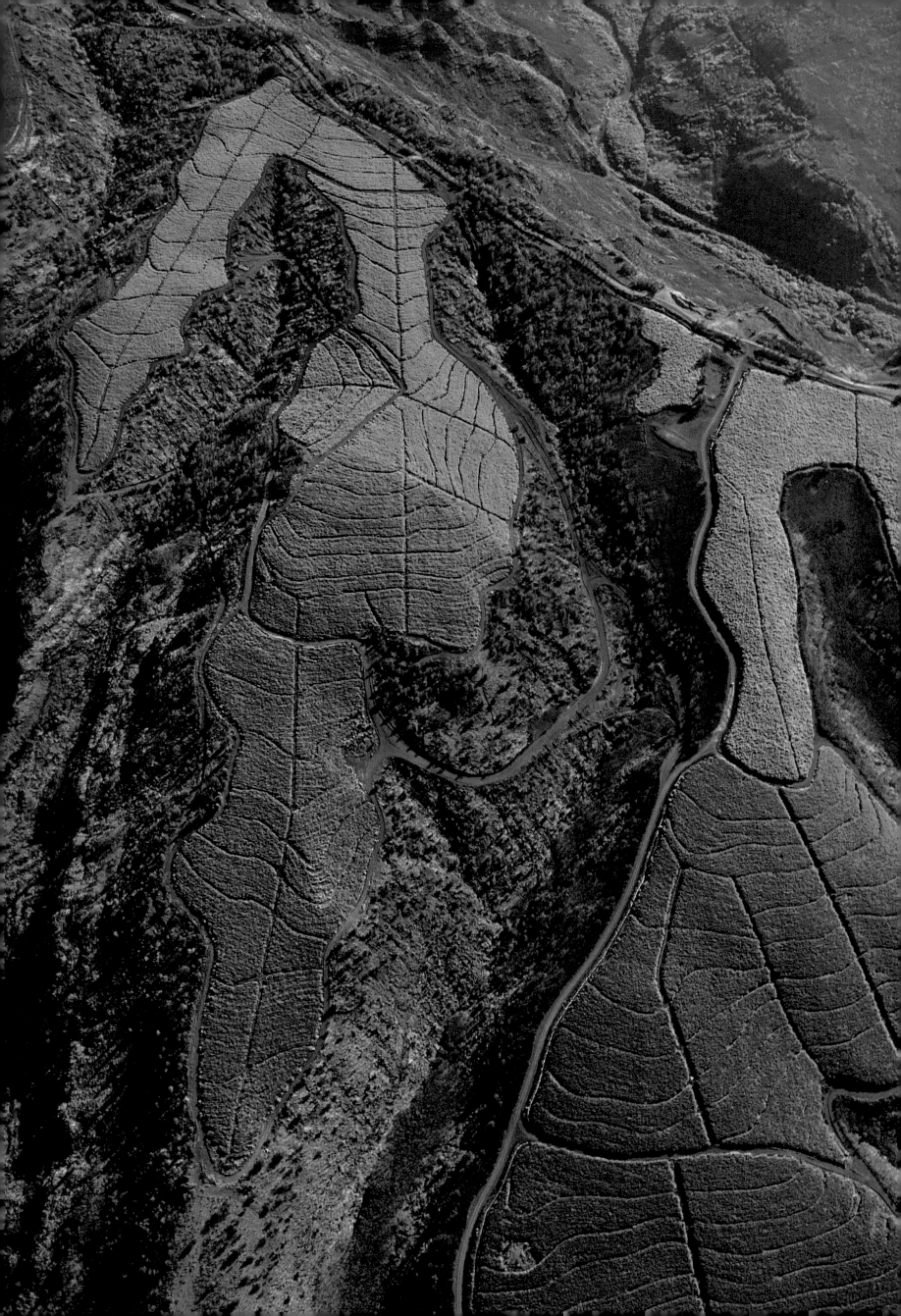

FIELD AND GARDEN

"No occupation is so delightful to me as the culture of the earth, and no culture comparable to that of the garden," wrote Thomas Jefferson from Monticello, his estate in Virginia, where he planted orchards, vineyards, vegetable gardens, berry patches, and flower beds. He experimented with nearly seventy species of vegetables, including such exotica of the day as eggplant, sea kale, potatoes and tomatoes. French artichokes lent an international flavor to his table, while Mandan corn, brought back by the Lewis and Clark expedition, added a taste of America's newest frontier. In a sunny nook, he planted peas as early as February, which would have given him an edge in the neighborhood race to produce the first English peas each spring. The winner had to host the losers at a dinner featuring the fresh-picked evidence.

Today, four out of five American households garden, cultivating vegetables, nurturing flowers. Truck farmers serve metropolitan areas, while far-flung large-scale growers supply a nationwide market. Sunshine and water in abundance permit cultivation of vegetables year-round in Florida and California, and, with expanded irrigation, even the arid Southwest yields out-of-season bounty. Lettuce and tomatoes account for 49 percent of the value of our fresh vegetable crop, onions 15 percent, and sweet corn, carrots, cauliflower, celery, and broccoli about 6 percent each. Wisconsin and Minnesota are prolific sources of vegetables for processing. Potatoes thrive in a swath from Maine to Washington.

Irrigation has changed the farming patterns of historic southern field crops: The cotton belt now stretches westward through Texas to California; sugarcane, once concentrated in Louisiana, flourishes in Florida, Hawaii and Texas. As the ranks of smokers have thinned, tobacco acreage has dropped sharply. Still, a grower can gross nearly $5,000 per acre in a good year, and tobacco remains very profitable. But not as profitable as another weed: Marijuana grown in a high-tech greenhouse covering a mere 1/100 of an acre – as it is in the "emerald triangle" of northern California – could gross as much as $5,000,000 annually. Legitimate greenhouse crops and landscaping products are big business in America. We spend nearly a billion dollars a year on potted plants and about three billion dollars on lawn maintenance.

Growers of fresh-market perishables that require hand harvesting rely on a pool of migratory farmworkers, mostly aliens based in Florida, Texas and California. In late spring, itinerant pickers flow north, following ripening crops through the summer and fall to the Canadian border. Low pay, poor living and working conditions, and job insecurity remain hallmarks of seasonal farmwork. Few Americans opt to do it, and growers fear that amnesty under the immigration reform program will encourage aliens to abandon field work for other jobs. The specter of a labor shortage gives new impetus to mechanization of the fruit and vegetable industry. But progress hinges on developing not only equipment but also crops – to ripen uniformly, to grow to a uniform size and shape, to withstand machine harvesting. Tomatoes have been bred for firmer flesh, tougher skins, and a boxy shape; picked green, they are turned red with ethylene gas. Iceberg lettuce has been bred to stay crisp for a three-week shelf life, but it stubbornly refuses to mature on schedule or square its curves for efficient packing.

What's good for growers, however, may not be so for other sectors. Far from bettering labor's lot, mechanization threatens farmworkers with loss of their livelihood. And consumers must face anemic-looking lettuce, devoid of flavor, and day-glo tomatoes that, as one critic put it, can be "drop-kicked from the produce section to the checkout counter." There is no panacea, but increasingly, discriminating consumers are willing to pay premium prices for premium produce, specialty growers are springing up to cater to them, and skilled farmworkers are finding secure employment. Still, nothing can rival home-grown beauties. Nothing can match sun-warmed tomatoes picked at noon and eaten at lunch, nor for dinner can anything beat the sweet satisfaction of the season's first peas.

Left: Fields of sugarcane finger their way along mountainous ridges on the island of Kauai in Hawaii. Horizontal terraces check erosion; a submain snaking alongside a road supplies water for drip irrigation. Despite difficult terrain, abundant sunlight and water bring a sweet harvest: Sugarcane yield tops 100 tons per acre in Hawaii, more than triple the yield on the U.S. mainland.

Overleaf: Pink petunias, orange and yellow marigolds pop out among the pansies and other bedding plants at a wholesale nursery in Carpinteria, California. Mexican pottery is stacked in the docking area. Satisfying Americans' itch to exercise their green thumbs is a multibillion-dollar business.

● *Top left*

HARVESTING ONIONS in upstate New York, seasonal farmworkers gather the crop into 40-bushel bins. The rich muck soils, which yield succulent yellow onions, derive from decomposed plants that settled in ancient swamps. Choked with weeds and algae, a ditch drains the fields of excess moisture. When onions reach the desired size and stage of maturity, they are topped to stop growth – once done by hand with shears, the operation is now mechanized – and left to cure in the field for about two weeks. They are then harvested and taken to storage facilities where near-freezing temperatures induce dormancy.

● *Bottom left*

ORLANDO GOLD and other carrots thrive in Florida's fertile muck soils – a.k.a. black gold. South of Lake Okeechobee, two harvesters spew carrots into tractor-drawn field wagons, which will deliver them direct to the packing house. Besides a market basket of winter vegetables – tomatoes, broccoli, carrots, celery, eggplant, lettuce – nearly half of the nation's sugarcane is grown on drained organic soils in southern Florida. Since the land is almost level, little soil is lost to erosion, but muck is highly vulnerable to other forms of degradation. Draining accelerates decomposition of the plant residues that make up the muck, and as carbon in the plant matter oxidizes and returns to the atmosphere, the soil subsides about an inch a year.

● *Top center*

BLUSHING A dozen shades of pink, sweet peas rub cheeks with white alyssum in California's Lompoc Valley, the world's largest producer of field-grown flower seed. Planted in varietal blocks, marigolds and lobelias, delphiniums and dahlias carpet the 12-mile-long valley, where the climate is tempered by breezes off the Pacific. After the flowers have set seed, the blocks are harvested.

● *Top right*

TOMATOES PLANTED in February get off to a fast start with center-pivot irrigation and early warm weather in southern Georgia. Heavy spring rains when cover is scant have caused erosion. Through May, millions of tomato, cabbage, pepper, other bare-root transplants are hand-harvested by migrant labor and trucked to northern states and Canada. There farmers on contract to canners bring the crops to maturity.

● *Bottom center*

DELICATE STITCHERY of seedlings, shrubs, and trees ready for transplant skips across contoured terraces at a nursery near Ottawa, Kansas. Always a symbolic act, planting a tree in prairie country made a haven in the wilderness. Arbor Day, instituted in Nebraska in 1872, turned planting trees into an annual American ritual.

● *Bottom right*

ALIGNED IN manicured hedgerows, tea bushes on Wadmalaw Island, South Carolina, trace their roots to plants imported from the Orient as early as 1800. Different greens mark different varieties; some 300 are grown on the Charleston Tea Plantation, which produces the only tea grown in America. Between May and October, a mechanical harvester periodically shears two or three inches of new growth off the top of the hedges. The custom-built clipper does the work of hundreds of pickers hand-plucking two leaves and a bud.

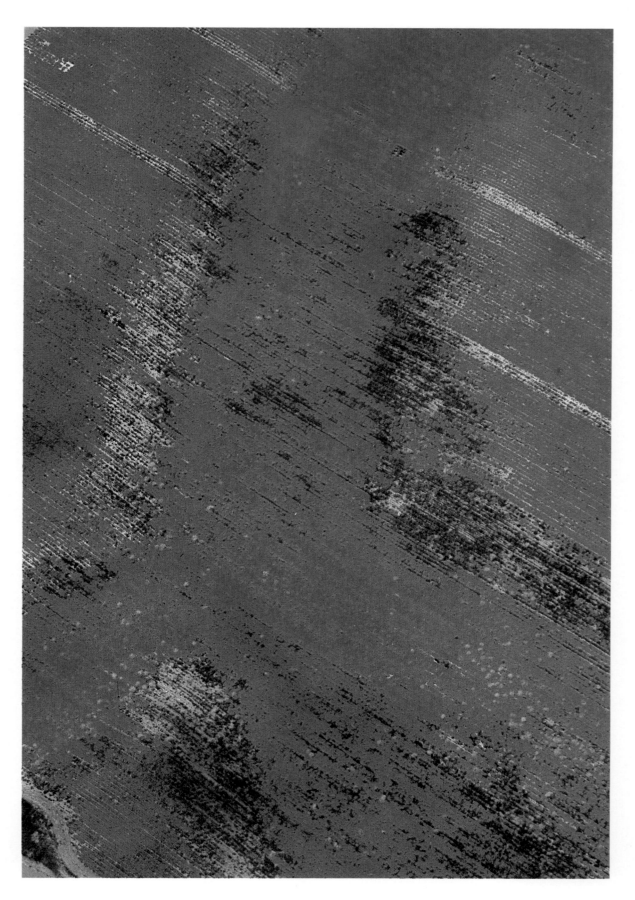

● *Above*

COLORIST CANVAS to some, a field of California poppies near Lompoc, California, looks like a crop failure to a seed grower. The likely culprit? Rain, causing the plants to produce leaves rather than species-perpetuating flowers and seeds.

● *Right*

CONTOURED PLANTINGS of shade trees conserve soil at a Missouri nursery. Washington Hawthorns and Pin Oaks wear fall foliage, while Sawtooth Oaks remain stubbornly green. Dark soils have been fall plowed for spring planting.

● *Right*

SHADED BY saran and shielded by trees, ornamental plants for the landscape and floral trade are cultivated under optimal conditions in southern Florida. By moderating light intensity, temperature, or evaporation rate, crop covers can enhance the natural advantages of the state's warm, humid climate. Exotic tropical crops, which figure prominently in the ornamentals industry, can be grown year-round in the peninsular south. The only serious threats to production are an occasional hurricane or an even rarer freeze.

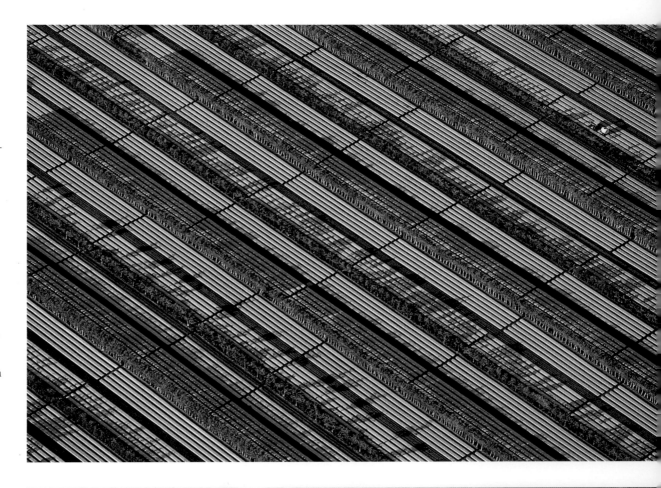

● *Right*

DARK SHEETS of translucent plastic filter out precisely 27 percent of central Florida's sunlight, simulating the shady groves in which ferns thrive naturally. Ferneries in Volusia County grow 85 percent of the fern used by U.S. florists, primarily the feathery leatherleaf fronds used to fill out arrangements of cut flowers. Like fruit and vegetables, floral fern is a labor-intensive crop. Workers, largely Mexican nationals, bend over the two-foot-high plants to carefully hand-cut the fronds, which grow back in six to eight weeks. Paid by the bunch of twenty-five fronds, some cutters can harvest as many as 400 bunches in a seven-hour day — almost a bunch a minute. The industry is highly seasonal. Seventy percent of the sales come in the first five months of the year — a period when millions of men choose to "say it with flowers" — with peaks at Valentine's Day in February and Mother's Day in May, and in between, Easter and National Secretaries Week.

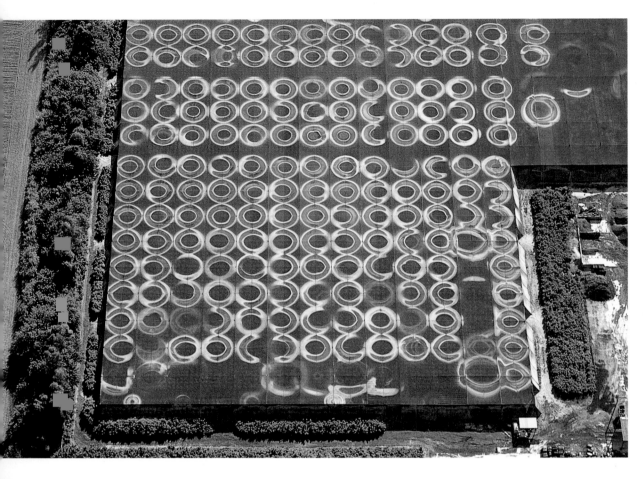

● *Left*

RESIDUE RINGS pattern a shade house covering two or three acres near Homestead, Florida. The mineral deposits result from a sprinkler system installed beneath the saran cover to water ornamental plants growing in containers. The cover, a woven polymer that lasts up to ten years, protects the plants from drying winds and reduces their need for water. More significantly, the dark saran acclimatizes foliage plants for use in low-light interiors — homes, offices, business, malls.

● *Left*

POLYPROPYLENE TENTS create a tropical climate in temperate Connecticut for growing tobacco ideal for cigar wrapping. The shade cloth traps humidity and shields tender plants from the sun in the Connecticut River Valley, much as low-hung clouds do in steamy Sumatra and Cuba, traditional centers for producing fine cigar wrappers. The results are comparable: aromatic leaves that are thin and elastic, uniform in color, free from blemishes, and small veined, so a cigar burns evenly and holds a long ash, and is pleasing to the eye as well as to the nose and palate. The tobacco is hand-harvested, primarily by seasonal workers from Jamaica, who pick the leaves as they ripen, usually the three bottom leaves per priming, or run, at weekly intervals from July to September. In the curing sheds, the leaves are stitched to laths by machine and hung to dry. They gradually turn from green to the pale golden brown that marks prime wrapper tobacco.

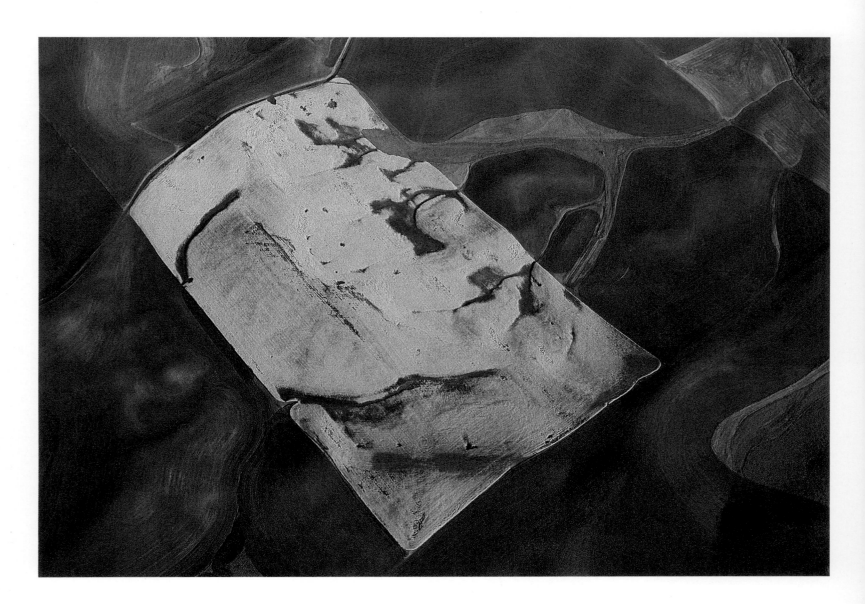

RAPE MANTLES a hilltop with spring bloom in Idaho's Palouse region. Seeded in July, rape provides excellent cover during the winter when erosion risk is greatest. In the U.S. rapeseed is used mainly for birdseed and lubricating oil, but new hybrids have spurred interest in it for low-cholesterol cooking oil. Among edible oils worldwide, rapeseed ranks fourth, after soybean, palm, and sunflower oil, which contain 15 percent, 51 percent and 11 percent saturated fat respectively, compared to 6.8 percent for rapeseed oil.

SURF OF yellow sunflowers ripples across the North Dakota prairie, home of 80 percent of the nation's crop. Well adapted to the northern plains, sunflowers resist drought and mature early. Sunflower seeds yield cooking and salad oil, and are used in confectionery, snacks, and livestock feed. The Indians ate sunflower seeds raw and cooked, as well as the roots of some species; the seeds also provided hair oil, the flowers dye, and the stalks fiber. For fall dance ceremonies, Hopi maidens use a powder made of sunflower petals, which gives their skin a golden sheen.

COHORTS OF Japanese yews – as many as 5,000 to the acre – march across the fields of a nursery in Rocky Hill, Connecticut. Thinning the ranks, tractor-drawn trailers haul away trees for shipment to market. The smaller yews are nine years of age; for customers demanding instant gardens, the taller ones are 14 years old. In the arid West, landscaping, often dominated by exotic plants from humid climates, accounts for 43 percent of home water use.

● *Above*

● *Right*

POTATO HARVEST in, a farmer in Aroostook County, Maine, prepares the fertile caribou loam for next season, leaving grassed waterways intact to control erosion. A third of Maine's crop is sold as seed potatoes, twenty to thirty varieties.

PULLED BY tractors through a soggy Maine field, bulk trucks keep pace with mechanical potato harvesters, which dig the potatoes, sift out soil, pick out residues, and presort stones from potatoes. Humans make the final separation.

● *Right*

MULTIGANG MOWER on a turf farm in Rhode Island cuts a 12- to 24-foot swath with each pass. At the peak of the season, the grass may be mowed every other day to promote growth and density. In the lower section, a sod harvester has shaved off 18-inch-by-6-foot strips, removing a mere ¹/₈ to ¹/₄ inch of loam with the sod. Almost all of the roots are left behind to help rebuild the soil. Buyers of sod include golf courses, cemeteries, condo complexes, homeowners — anyone who wants a ready-made carpet of living grass. Turf grass covers some 50,000 square miles in America, but most people grow their own, alternately force feeding and clear cutting an area the size of Alabama in a weekend ritual.

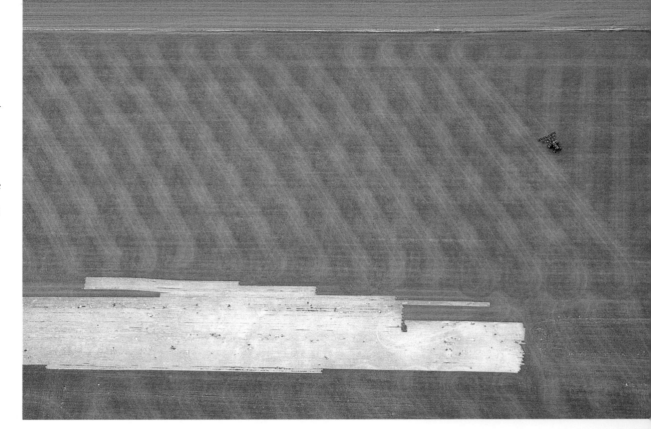

● *Right*

CHOOSING AMONG a gamut of greens in a display yard, a tractor driver fills an order at a wholesale nursery in southern New Jersey, where ornamental shrubs and flowering plants are ready for shipping. The tradition of gardening in America goes back to farmwives who brought flower seeds to the New World more for medicinal than for decorative use. All-purpose marigolds removed warts, cured toothache, dyed hair. Avid gardener Thomas Jefferson passed his passion on to his grand-daughters. Writing to one who had helped plant bulbs in the fall, he described the fruit of their labors: "The flowers come forth like the belles of the day, have their short reign of beauty and splendor, and retire, like them, to the more interesting office of reproducing their like." The glory of spring maturing into the fullness of summer, the ripe harvests of fall — these recurring wonders still bind one generation to another, and one gardener to another, from backyard horticulturists to commercial growers.

● Left

OBSTACLE COURSE for tractor drivers? No, sorghum planted for food and cover for pheasants on a shooting preserve in New Jersey. The pattern was mowed during the growing season; come fall, pheasants being hunted will flush at the breaks, unable to run to the end of a strip under cover. New Jersey, traditional supplier of farm produce to the New York metropolitan area, now caters to other urban tastes. The biggest agricultural sector, greenhouse and nursery products, keeps growing along with the suburban sprawl, but dairy cows are being replaced by dressage horses. Since 1961 the horse population has more than tripled, while the dairy herd has dwindled to less than a quarter of its former size.

● Left

PUMPKINS, CUT from the vine and waiting in line, will be hauled by tractor-drawn trailer to swell the piles at the end of a field near Fresno, California. Bare spots appear where plants have been lost, presumably to mosiac, a viral disease that can wipe out a harvest. More than 95 percent of the crop will be carved into Halloween jack-o-lanterns. Today, pumpkins range in size from tiny to gigantic. Jack Be Little, a two-inch-high quarter-pounder, adds color to a harvest table. Big Moon, nearly three feet tall and 200 pounds, is a favored contender for blue ribbons. Pumpkins belong to the gourd family of cucurbits, which includes squashes, cucumbers, and watermelons as well as ornamental and utilitarian gourds. Archeological evidence suggests that Indians grew gourds – both bottlegourds and the same species used to make jack-o'-lanterns – at sites in Illinois some 7,000 years ago, which would make them the first plants to be cultivated north of Mexico.

● *Above*

PARALLEL BEDS of tobacco fields
echo the rolling contours of North
Carolina's upper coastal plain.
Green fields are small grain that
was planted in the fall and is start-
ing to tiller. Tobacco transplants are
set out in the spring on raised beds
to help keep their leaves off the
ground and their roots well
drained. Every seventh row is
skipped to leave a "sled row" for
equipment. Tobacco production is
concentrated in Appalachia, flue-
cured tobacco in North Carolina
and aromatic burley in Kentucky
and Tennessee, where seedlings
are still set out by hand and ripe
stalks hand-harvested.

● *Facing page*

RIBBED WITH roads and veined with
crop rows, tomato fields straddle
the hills north of San Diego,
California. Such steep terrain –
the only land available for agriculture in the coastal zone, prized
for its long growing season – is
costly to farm. To combat erosion
the rows have been contoured.
To maximize production, these
fresh-market tomatoes are "pole
grown," drip irrigated, and heavily
fertilized. From May through
January, migrant workers hand-
harvest the tomatoes, picking over
the fields as often as three times
per week. The payoff is annual
yields of 30 to 50 tons per acre.

● *Overleaf*

LAPPED BY the blue Pacific, a sea of
green sugarcane surrounds Numila,
the plantation hub, on the island
of Kauai in Hawaii. The greenest
fields are young cane, still growing;
as the cane ripens, the fields yellow;
the brown fields have been cleared
or just replanted, normally done
after five cuttings. The fields are
drip irrigated, for cane is a thirsty
crop, using a ton of water to
produce each pound of raw sugar
before extraction. Lying like warts
on the landscape, piles of rocks
gathered from the fields when
labor was cheap remain there
today, too costly to move.

FRUIT AND NUTS

"God alone can improve apples," preached Johnny Appleseed, frontier apostle of American orchards. Around 1800 he loaded a canoe with apple seeds collected from cider presses in Pennsylvania and paddled down the Ohio and into the wilderness. At intervals he would clear a patch of ground, girdle the trees, and sow his seeds broadcast. Over the next half century he was intermittently sighted leading a seed-laden mare over forest trails, himself on foot – usually described as barefoot and wearing a coffee-sack shirt and a mushpot hat – or drifting downstream in a dugout overflowing with saplings. For he peddled apple trees to settlers and, with missionary zeal, gave them away to those who would or could not pay. But he was no "pauper-philanthropist," as legend states; he was a Yankee entrepreneur with an eye on the market. Born John Chapman in New England, he started out as a squatter, but over his career he established more than thirty nurseries as his plantings leapfrogged westward one jump ahead of the farmers. As Hovey's *Magazine of Horticulture* explained in 1846, "When the settlers began to flock in . . . old Appleseed was ready for them with his young trees; and it was not his fault if everyone of them had not an orchard planted out and growing without delay."

Planting trees means putting down roots. In 1539 Hernando de Soto laid claim to Florida by planting orange trees around his camps; in 1623 the first apple orchard was planted in New England; and in the 1770s Franciscan missionaries planted vineyards in California. These areas are still major fruit-growing regions. Bathed by warm, humid air from the Gulf and the Caribbean, Florida produces about 70 percent of the nation's oranges and 80 percent of the grapefruit. Thin-skinned and juicy, Florida's oranges are ideal for processing. Virtually all fresh-market oranges and most of our lemons are grown in California, where wider daily temperature swings make citrus brighter in color, higher in both sugar and acid, and thicker-skinned – hence cushioned against shocks of shipping. Hard winter frosts and high summer rainfall make the Northeast prime apple country; Washington, however, where irrigation offsets dry summers, ranks first in apple production. Blessed with richly varied terrain and climate, California leads the nation in production of about fifty crops, including strawberries, kiwifruit, avocados, almonds, apricots, and both wine and table grapes. In tropical Hawaii and southern Florida, many ethnic communities carry on the American tradition of nurturing immigrant plants: pungent orange papayas, star-shaped carambolas, ruddy-cheeked mangoes, dusky purple passion fruit.

Since crops were first cultivated, farmers have battled pests that prey on them. In recent decades the agricultural industry has waged all-out chemical warfare on insects, fungi, disease, and weeds – only to reach a standoff: Although U.S. farmers apply more than 400,000 tons of pesticides to fields and orchards annually, pests still destroy about a third of the crops. Other chemicals are enlisted to enhance color, ensure crispness, extend storage life. Intensive use of agricultural chemicals has opened a Pandora's box of untoward consequences. Developing apace, insecticide-resistant insects number 447 species by one recent count. Annually, an estimated 300,000 farmers and farmworkers suffer illnesses attributed to on-the-job exposure to toxic chemicals. Pesticides have polluted groundwater, and toxic residues have insinuated themselves into the nation's food supply, posing health risks for consumers. The end of innocence comes with the knowledge that of all the fruits of the garden the most chemical-laden are apples.

Many farmers argue that cutting back on chemicals would sharply reduce yields, but a growing number now favor integrated pest management, in which judicious application of pesticides is combined with other methods of control such as crop rotation and the introduction of natural enemies of crop pests. The ideal solution to pest predation is, of course, pest-resistant crops. Gene transplantation now holds out the promise of crop plants that will be unappetizing to insects and immune to viruses. We cannot return to Eden, but we can come much closer to a pesticide-free environment.

Left: Corralling ripe cranberries, workers raft their buoyant harvest to the bank of a flooded bog near Plymouth, Massachusetts. There the berries are sucked up by a hose and run through a dechaffer, which siphons off leaves and other debris. A conveyor then hoists the crimson bounty into a semitrailer with a capacity of 40,000 pounds of berries.

Overleaf: Spectacular sea of cranberries is girdled by floating booms as wet harvesting gets underway in southeastern Massachusetts, where growers have added sand to the acid peat soil and controlled the flow of water through natural marshes to provide an ideal environment for this native American fruit.

● *Above*

A MAZE of interlocking crooks and islets, citrus groves on the east coast of central Florida occupy land that was once forested and pocked with ponds and sloughs. Excavation of marshy areas with draglines provided fill to build up ground for planting beds. Today dredge-and-fill reclamation is rare, thanks to wetland protection.

Oranges, of Asian origin, were brought to America by Columbus in 1493. Early oranges were bitter, but sweet oranges, grafted onto bitter-orange rootstock by Spanish missionaries, were flourishing in Florida by 1650. Modern growers prefer disease-resistant rough lemon rootstock for orange trees.

Grafting bud-bearing branches onto hardy rootstock has been practiced since Biblical times, and most of the trees in America's commercial orchards – 50 million orange trees, 30 million peach trees, 70 million apple trees – are grafted, notwithstanding Johnny Appleseed, who reportedly protested, "It is wicked to cut up trees that way." Propagation by grafting, however, may soon be superseded by tissue culturing in petri dishes, which can produce thousands of disease-free clones in just a few months.

● *Right*

CONCENTRIC ROWS of citrus trees wheel round the terraces of an orchard near Harlingen, Texas. The land has been bench leveled on the contour for efficient irrigation and runoff control, as has the green field, probably planted in perennial grass. The latest method of leveling land for irrigation employs high-tech laser grade control but usually produces large rectangular benches or fields, which, from a visual perspective, lack kinetic energy.

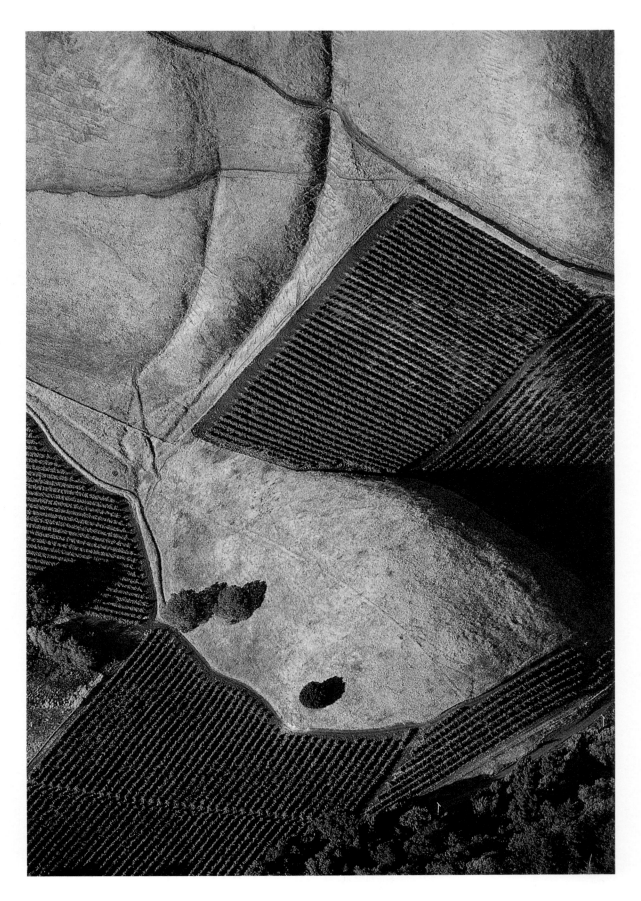

● *Above*

VINEYARDS CLIMB the slopes of narrow
Napa Valley, premier wine region
in northern California. Some 40
states boast wineries, but California
accounts for 90 percent of U.S.
production, from inexpensive jug
wines to world-class vintages.

● *Right*

NATIVE OAKS dot the Firestone Vine-
yard in the Santa Ynez Valley, a
young winery in an area where
viniculture is as old as the Spanish
missions. Reservoirs provide
backup water for frost protection
and irrigation.

106

● *Top left*

POLYESTER SWADDLING ensures that
locally grown, vine-ripened
cantaloupes and watermelons will
be ready for sale at the farmers'
market in Kennedyville, Maryland,
in time for Fourth of July picnics.
To meet this date, the grower sets
out transplants that have gotten a
head start in his own nursery, then
covers them with polyester fabric,
rolling it out by hand and
anchoring the edges with dirt so
they can be lifted for venting as
necessary. Lying lightly on the
melon plants, the porous sheeting
lets rain and sunshine through but
traps heat, acting as an incubator
to speed growth and ripening.
Sometimes the fabric can be
reused, but a stormy spring left
these strips in tatters.

● *Bottom left*

A CLUTCH of hives at the edge of an
almond orchard in Cache Creek
valley in northern California shows
honeybees are on the job for the
blooming season. Because all
major almond varieties require
cross-pollination, growers contract
with beekeepers to bring in
colonies of migrant workers to
perform that task – two or more
hives of strong, active bees per
acre, set out in clusters of six to
twelve. Each year over 200,000
hives are shipped to California for
almond pollination from out of
state, mostly from areas where
bees cannot survive the cold
winters. California weather is
generally mild, but almond trees
are early bloomers – through
February into March – and cool or
inclement spells even above
freezing can cut yields.

● *Top center*

LADEN WITH cranberries, a truck
backs onto a lift at a receiving
station in Carver, Massachusetts.
When the lift tilts, 18 million
berries will cascade out and pass
through cleansing blowers before
dropping into hoppers. The
Indians combined cranberries and
venison paste to make pemmican,
which they used as a trail mix.
Today Ocean Spray, a growers'
cooperative, has become a Fortune
500 company with an ever-
expanding line of cranberry
products.

● *Top right*

EXPLOSION OF ephemeral white
blossoms in southwestern
Michigan will metamorphose into
an abundant crop of tart red
cherries to go into pies and tarts,
jams and jellies, ice cream and
yogurt. Trees in the younger
orchard will not be ready to
harvest for five years or so. Until
then, they will be chemically
inhibited from flowering and
fruiting to encourage growth.

● *Bottom center*

HILLS OF almond hulls heaped up
by a custom huller in California's
San Joaquin Valley provide high-
fiber feed for livestock. The
almond shells are burned in
cogeneration plants, which supply
electricity to homes and industry.
The kernels, packed in round bins,
are shipped to processors and
packagers. Almonds were once
harvested by striking the trees
with wooden mallets and
gathering the nuts in a tarp. Today
97 percent of the U.S. crop –
virtually all grown in California –
is harvested with mechanized
shakers and sweepers.

● *Bottom right*

SUN GOLD of cottonwoods eclipses
subtler fall hues of apple trees in
the Chama Valley in New Mexico.
Whatever the variety – McIntosh,
Rome, or Granny Smith –
connoisseurs claim that apples
grown in irrigated orchards tucked
away in the state's northern valleys
derive a superb flavor from the
pure mountain water.

● *Above*

SUFFUSING A drift of peach blossoms, dawn nuzzles a farmstead to life on the western rim of the Michigan Peninsula. Famed as a fruit belt, the region owes its benign climate to Lake Michigan, which, in conjunction with prevailing winds, acts as a giant heat pump, moderating temperatures year-round. In the spring, this large body of cold water cools warm air crossing it, delaying bloom and hence lessening the risk of damage by spring frost. By fall the lake has become a relatively warm body of water, which warms cold air crossing it, cushioning sharp drops in temperature.

● *Facing page*

COUNTERPOINT OF tree and shadow animates an almond orchard in the San Joaquin Valley in California. Planting patterns seek the fine line between maximum number of trees per acre and crowding, which cuts into yields as trees mature. Setting trees 30 feet apart on a square grid, growers can plant forty-eight trees per acre; by spacing trees 25 feet apart on a hexagonal grid, they can plant eighty trees per acre. For cross-pollination, most almond orchards comprise two or three varieties, although some consist of trees of one variety with limbs of a pollenizing variety that have been budded or grafted.

● *Overleaf*

CONTOURED WITH cartographic precision, rows of avocado trees parallel diversion channels installed to control erosion in the foothills of California's Santa Ynez Mountains. Avocados like the well-drained soils of these uplands, but on such steep slopes the erosion potential is high, as are orchard operating costs. In southern California, however, urbanization of more gently sloping areas often leaves no other option. Native to Meso-america, avocados were cultivated by the Aztecs. Other subtropical fruits grown in the state include guavas, figs, dates, cherimoyas, papayas, and citrus fruits.

● *Above*

PLANTING PECAN trees in clay soil
near Las Cruces, New Mexico, a
backhoe has dug pits 10 feet deep
to reach sand. Before the holes are
refilled, sand and soil will be mixed
to improve drainage and keep the
trees from "drowning."

● *Right*

DATE PALMS tracing their origin to
Saharan oases thrive in hot, dry
Coachella Valley in southern
California. As fruit clusters ripen,
they are enclosed in paper bags
to protect them from insects and
birds, dust and sunburn.

● *Above*

STRAWBERRY FIELDS forever ride a
sea of slitted plastic near Carlsbad,
California. Warming the soil and
conserving water, the polyethylene
mulch promotes development. Like
deciduous fruit trees, strawberries
need chilling to store up energy
that later explodes in a profusion
of fruit. Transplants that have
been exposed to cold in northern
California nurseries are set out
in the fall in southern California,
where it is warm enough for them
to grow in winter. The harvest runs
from late January to July, yielding
30 tons of strawberries per acre.

● *Right*

LAYOUT OF a pineapple plantation
on the island of Lanai in Hawaii
echoes agribusiness efficiency.
Unspooling from irrigation ditches
and access arteries, service roads
scroll around each field. The width
of a field is twice the length of
the booms that carry the nozzle
assemblies of spray rigs and the
conveyor belts of harvesters. A
field's rounded ends match the
turning radius of the equipment.
Today just eighteen highly
industrialized farms produce
virtually all of Hawaii's crop.
    A native of the Amazon basin,
the pineapple became the emblem
of Hawaii with a boost from Yankee
promoter J.D. Dole, who persuaded
homesteaders in the territory to
plant "pines" and pioneered mechanized cultivation and canning in
the early 1900s. In 1922 he bought
the entire island of Lanai and
established a 14,000-acre plantation.
When Dole died in 1958, the
Hawaiian Pineapple Company,
which grew out of the cannery
he founded in 1903, accounted for
72 percent of world output. But
rising labor and land costs in
Hawaii turned the tide, and by the
mid-1980s the Far East was producing 70 percent of the world
supply of canned pineapple.

EARTH, AIR, FIRE, WATER
– AND THE HAND OF MAN

"The soile is fat and lustie," exulted one New World explorer. Others praised "the wholesomeness of the climate" and, tempering the sun's warm rays, "soft rain, which falls in such abundance." Though lacking most of the biological resources US farmers would draw on for crops and livestock, North America was singularly favored with the physical resources necessary for agriculture, and indeed for life: earth, air, fire and water. Green plants stretch skyward with their leaves to capture the sun's radiant energy and delve into the earth with their roots to feed on nutrients dissolved by water. Terrestrial plants nourish the herbivores, which in turn nourish the carnivores; both plants and animals provide sustenance for humans and other omnivores. Eventually all organic matter becomes fodder for decay organisms, which reduce it to inorganic forms that can again be taken up by plants.

Through agriculture man manipulates this natural cycle, diverting the flow of nutrients to plants and animals selected for their usefulness to humans. A uniquely hospitable environment – in terms of soils and topography, climate and latitude, streamflow and groundwater – has enabled farmers to grow almost anything in America, somewhere, and to do so in undreamed-of abundance. Only one-fifth of the earths land surface is suitable for agriculture, but more than half of the 2.3-billion-acre land area of the United States is used to produce crops and livestock. Alaska extends into the Arctic and Hawaii lies in the tropics, while the contiguous forty-eight states span a continent in the temperate middle latitudes, where the distribution of sunlight is ideal for growing most major food crops.

Solar radiation, the ultimate source of energy for virtually all life on earth, is used by plants in photosynthesis. This fundamental process converts inorganic nutrients – carbon dioxide and water – into energy-rich organic compounds – mainly carbohydrates – and releases free oxygen. Both plants and the animals that consume them unlock the energy by respiration – the converse of photosynthesis – in which carbohydrates and oxygen are recombined to yield carbon dioxide and water again. The most important energy input in agriculture is solar energy, but only about one part in 200,000 of the sunlight falling on earth is channeled into food for humans. No shortage of solar energy limits agricultural production.

Nor is there any scarcity of carbon dioxde or oxygen, which are readily available in the atmosphere. Nitrogen, another vital nutrient, makes up 79 percent of the atmosphere; a column of air over an acre of ground contains 75 million pounds of it. Yet plant growth is limited more by lack of nitrogen than by any other nutritional deficiency. Atmospheric nitrogen is useless to most organisms; it must first be "fixed" – or combined with other elements – before it can be utilized by plants. On land, nitrogen is fixed mainly by bacteria that live symbiotically in the root nodules of legumes, such as beans and alfalfa. Intervention in the nitrogen cycle, by cultivation of leguminous crops and production of chemical fertilizer, has led to more nitrogen now being fixed by these two methods than was fixed by terrestrial microorganisms before the advent of agriculture. While carbon, hydrogen, and oxygen are obtainable from air and water, nitrogen and the score of other elements essential to plants must be absorbed from the soil. Fertile soil is generated very slowly by weathering of rock, crystallization of minerals, accumulation of humus. A crucial component is the plethora of living organisms that populate healthy soil – as many as three tons of insects, worms, fungi, algae, bacteria, and other microbes per acre-foot.

Although water is the most abundant substance in the biosphere, a scarcity of water often limits plant growth. More than 97 percent of all water consists of salt water in the world's seas and oceans, and 2 percent is frozen solid in glaciers and the polar ice caps. Less than 1 percent is fresh liquid water, surface and subsurface, including 0.005 percent held in the soil within reach of plant roots. A mere 0.001 percent is water vapor, continually cycled between the earth and the atmosphere. Over the conterminous United States, precipitation averages 4,200 billion gallons per day, while evapotranspiration averages less than two-thirds of that amount. The difference lies in runoff, which feeds lakes and rivers and replenishes underground water, restoring soil moisture first then percolating down to refill aquifers. By exploiting runoff –

Left: Shaping the land to their liking, American farmers have converted huge areas of forest and grassland to the production of plants of their own choosing, such as the corn, small grain, and hay growing here in central Ohio. Planting the crops on contour strips and keeping steeper slopes wooded help prevent soil erosion on the hilly topography, while crop rotation preserves soil fertility

119

diverting surface flow or tapping groundwater for irrigation – farmers have reaped spectacular yields from land that would otherwise be marginally productive or even barren because of insufficient rainfall.

In natural ecosystems, nutrients flow in a closed cycle from the soil through plants, animals, and decomposers back to the nutrient pool in the soil. Fauna and flora reach an equilibrium with the resources available in the environment. With the beginning of agriculture, people began to disturb this equilibrium. Farmers devoted more and more resources – land, water, labor – to nurturing a few favored species, primarily cereals, legumes, and tubers, some fruits and vegetables, and ruminants, swine and poultry. At the same time, harvesting depleted the soil of nutrients by taking plants and animals out of the loop. Farmers maintained soil fertility by adding manure and other organic fertilizers. New England settlers caught fish in spring spawning runs and buried them in their corn hills to ensure a bountiful harvest. In the 1870s the systematic slaughter of the buffalo on the Great Plains yielded a bonanza of bones, which were used to revitalize worn-out eastern soils.

Since the mid-twentieth century, intensive cropping and monoculture have drawn down the working capital of the soil at an accelerated rate, and farmers have applied chemical fertilizers in escalating amounts to expand output. But this strategy may be approaching its limits. Some crops seem to be getting all the nutrients they can use, and giving them more no longer enhances yields. In addition, runoff laden with nitrogen compounds and other farm chemicals has become a major source of water pollution. Agriculture aimed at maximum yield also requires ample moisture, and irrigated acreage now accounts for almost one-seventh of our farmland and more than one-fourth of farm production. But irrigation also appears to be cresting. Demands on streams and rivers often exceed supplies, and profligate pumping of groundwater has caused water tables to drop, sometimes precipitously. As water supplies have declined, irrigation costs have risen, as has competition for water.

The primary energy source for modern agricultural technology is fossil fuels. Like respiration, combustion of these organic deposits unlocks stored solar energy and releases carbon dioxide as a byproduct. Natural gas is used to synthesize nitrogen fertilizer, while oil provides mechanical power to till the soil, pump water for irrigation, and plant and harvest crops. Today energy input per acre in terms of fuel often far exceeds energy output in terms of food. Moreover, fossil fuels have supplied the energy to squander land and water resources by overplowing and overpumping. In many areas we are losing topsoil faster than it forms and draining aquifers faster than they fill. America is richly endowed with resources for agriculture, but future productivity depends on how wisely we manage those resources.

Competing for land and water, industrial and urban development also threaten agriculture by fouling the atmosphere: Acid rain and air pollution are jeopardizing crop yields; chemical pollutants are eroding the ozone layer, which shields life on earth from harmful ultraviolet rays; and now man-made climate change may be adding to the hazards of natural climate variability. By trapping heat radiated from earth, the atmosphere warms and insulates our planet. Mounting evidence indicates, however, that an atmospheric buildup of carbon dioxide and other industrial gases is intensifying this greenhouse effect. By the year 2030 global temperatures could rise 3 to 9 degrees Fahrenheit – or as much as the earth has warmed since the last ice age. Possible impacts on agriculture include violent wind storms, disrupted rainfall patterns, increased frequency of drought, and loss of cropland due to rising sea levels as polar ice and glaciers melt.

Man's technological genius, which has so profoundly improved the human condition, may now be endangering human prospects by ominous interventions in the natural flow of nutrients and energy in the biosphere. A mere four-hundredth of the earth's radius in thickness, our home planet's biosphere – a tenuous membrane of air, soil and water aswarm with life fueled by energy from the sun – is utterly unique, so far as we know, the most beautiful, beguiling, and astonishing creation in the cosmos. The almost infinite variety of life on earth is dazzling, but even more marvelous are the similarities of form and function of living organisms, their conformity to a pattern of interlocking cycles. All are dependent on other species for survival; all belong to a symbiotic community sharing the same environment, not least of all humans as well as the plants they cultivate and the animals they husband. Man versus nature can only be a losing battle; we are part of nature – to win would be to doom ourselves. But if we honor our bonds with nature, as future generations of farmers plant their crops, whether last year be good or bad, they will be able to look forward to harvest, as farmers always have, with the hope that this year will be better.

Right: Early spring growth veils low rolling hills in Washington's Palouse country. The dark ribbon is the trail of a seeder that failed to engage, leaving blank rows that did not show up until the crop sprouted. Spring snowmelt can cause high losses of the rich loess soil on bare slopes. Extensive production of winter wheat in the region provides protective cover at this time of year.

Overleaf: At the peak of spring bloom, tart cherry trees flourish in the sandy soils and moderate climate of southwestern Michigan. The grower pulling an air-blast sprayer is not applying pesticide, which would damage bees during pollination. Rather, he is spraying after a rain to control brown rot, a fungus that can cause the loss of an entire crop, almost overnight in warm, wet weather.

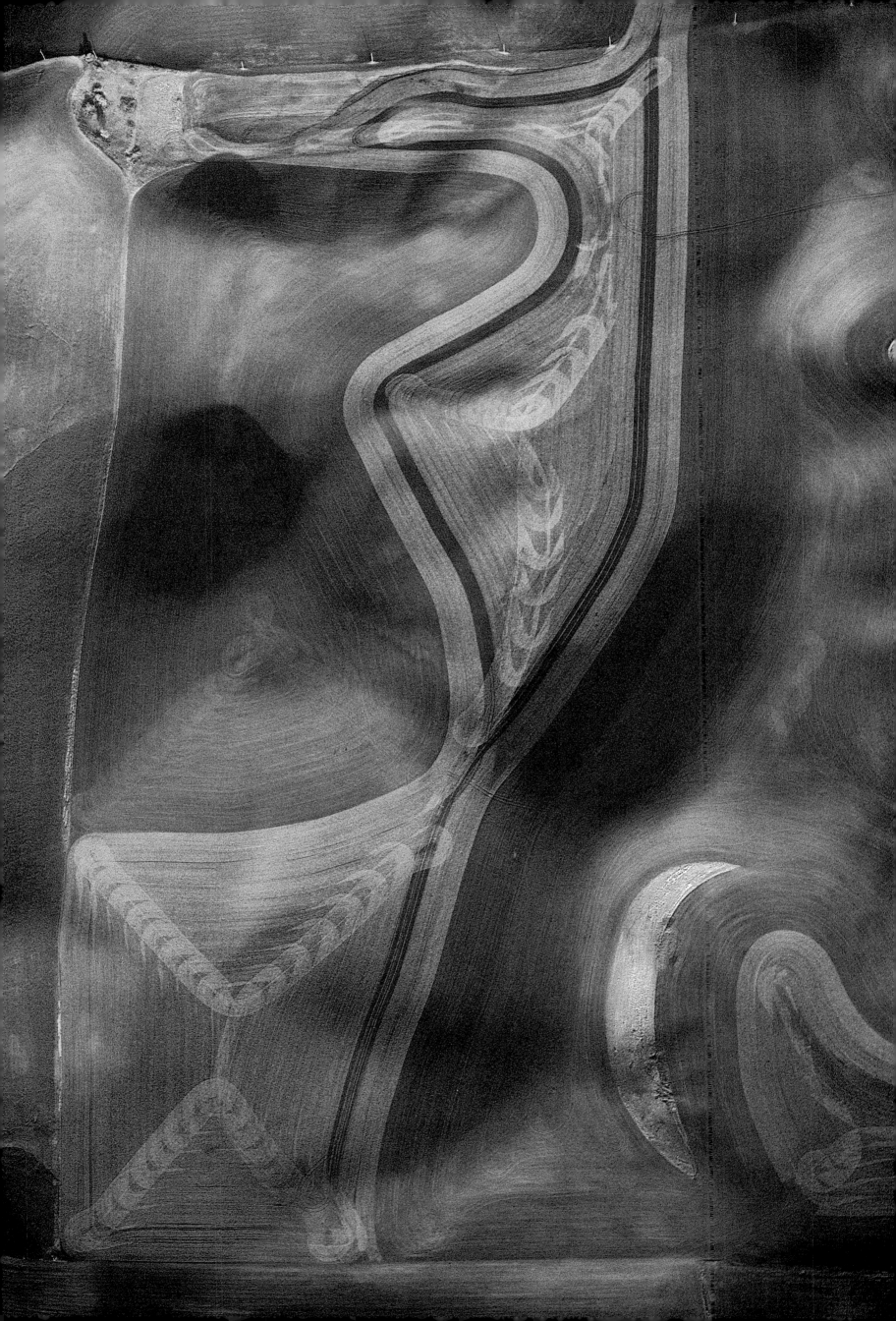

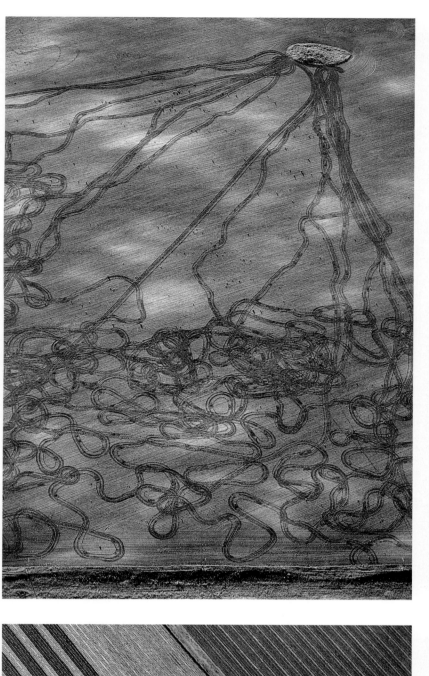

● *Top left*

TRAIL OF a rockhound near Pillsbury, North Dakota, was made not by a geologist but by a farmer who seems to have taken the opportunity during summer fallow to go over his field on a tractor with a rock-picker attachment and gather his trove in a central heap. Fewer stones will leave more space for growing next year's wheat crop. The rocks are grist from the glaciers that ground their way southward over the upper Midwest four times during the ice ages, leaving rock debris as they retreated.

● *Top center*

FOSSIL FISHBONE being unearthed near Sunnyside, in Washington's Yakima Valley, is actually a stage in the transformation of an arid mountainside into an orchard. A ridge running down Snipes Mountain is being smoothed by a bulldozer knocking off the top of the ridge and filling in the draws. Leveling will facilitate the use of sprinkler irrigation, which is essential for agriculture in this region, where rainfall averages only seven or eight inches a year.

● *Top right*

UPROOTED BY bulldozers, scrub pines stacked like cordwood will be hauled away to clear the ground for advancing rows of blueberries, near Hammonton, New Jersey. Blueberries are one of the few native U.S. crops of economic importance today; others include cranberries, pecans, sunflowers, wild rice, and Concord grapes.

● *Bottom left*

PULLING A four-row cultivator, a farmer cuts through the dark soil between rows of newly sprouted corn or soybeans to control weeds, near Spink, South Dakota. After finishing the first pass through the field, he will double back and cultivate the untilled strips. A lone tree, probably a hackberry, spills a puddle of shade on the flat, linear landscape. Willa Cather wrote of the vast prairie: "Trees were so rare in that country, and they had to make such a hard fight to grow, that we used to feel anxious about them, and visit them as if they were persons." Settlers often planted fruit trees around the farmstead, in handy orchards, near the kitchen door, or – a favorite spot – on the site of old privies as old holes filled and the farmer dug new ones.

● *Bottom center*

PREPARING a field for spring planting, probably to soybeans in late May, a farmer near Lawrence, Kansas, has spread fertilizer from a truck, leaving widely spaced vertical tracks. Now he is crossing those tracks, working the field diagonally with a tractor-drawn disk.

● *Bottom right*

FOUR-WHEEL-DRIVE TRACTOR pulls a 24-foot tandem disk and drag through highly productive muckland south of Florida's Lake Okeechobee. Spreading the weight of the heavy equipment, dual wheels help prevent compaction of the drained organic soils, a one-time swamp bottom that can give underfoot like a cushion. After the disk turns the topsoil, the drag crushes clods and levels the soil for planting sugarcane. The green crop in the next field is young sugarcane about eight inches high, and beyond is brown sugarcane stubble, the residue of last year's crop.

● *Left*

WHITE PLUMES trail out behind two spreaders applying lime to a mountain meadow on a windy October day, near Upperville, Virginia. Made up of any of several calcium compunds, agricultural lime reduces soil acidity, improves soil structure, and provides calcium and magnesium, essential elements for plant growth. According to their needs, farmers can alter the natural characteristics of soils in numerous ways, modifying soil alkalinity as well as acidity, draining wet soils, irrigating dry soils, and adding fertilizer to soils deficient in nutrients.

● *Above*

LIKE GOLDEN buckskin, brushed and fringed, ripe peas and harvest residue mantle the undulating hills of the Palouse region, near Pullman, Washington. Separating the peafield from dark summer fallow, the lines follow the land's contours, as do the rows of crop residue left in the wake of the harvesting combine. The Palouse harbors some of the most fertile, and most erodible, soils in the nation. Contour cultivation and crop residue management are two of the many techniques local farmers use to control erosion.

● *Overleaf*

AMBER FIELDS of corn and green alfalfa hay and pasture grass paint an autumn panorama on the rolling terrain of southwestern Wisconsin. Some corn has been cut for silage, but most still awaits harvest for grain. The alfalfa yields three or four cuttings for hay during the growing season. These crops, along with bluegrass pasture, nourish cows for dairy operations on numerous small family farms, which help keep Wisconsin first in the nation in milk production despite increasing competition from factory farms.

● *Above*

CUTTING IN concert, six combines
equipped with 24-foot headers
harvest a 144-foot swath of barley
on 22,000-acre Kolstad Farms in
Liberty County, Montana. The 165-
horsepower behemoths cut the
standing grain, thresh it, and
separate the grain from the chaff;
the chaff is chopped and blown out
the rear, and the grain goes into a
holding tank to be unloaded into a
truck on the run. Once grain is
ready for harvest, growers want it
binned as soon as possible. Hail,
wind, or wildfire can cause major
losses. Even rain – almost always
welcome in Montana – can delay
harvest, since grain must be dry
for storage. And rain can cause

barley to sprout, which means it
must be sold as feed grain rather
than to maltsters at a premium.
    The harvesting crew, out of
Kansas, is one of about 3,500
custom cutter outfits that travel
north with the ripening grain from
Texas to Canada, harvesting for
hire. Armed with state-of-the-art
combines costing as much as
$100,000 – which few farmers can
afford to spend on equipment that
sits in the shed for fifty weeks a
year – custom cutters make quick
work of a harvest, putting in 10-,
12-, even 16-hour days, sometimes
switching on the headlights and
working through the night. They
are paid by the bushel, so when

drought cuts yields, they suffer as
much as the farmers – or more,
since neither crop insurance nor
disaster relief helps offset their
losses. In 1988, when this picture
was taken, this field produced only
thirteen bushels per acre, less than
a third of its usual yield. Most malt
barley is irrigated, but the Kolstads,
who have been farming in
Montana since 1910, grow it on
dryland, gambling that it will meet
brewers' standards for plumpness.
They hedge their bets with other
grains, however, mainly winter
wheat, and – judging from the size
of their spread – have done so very
succesfully.

● *Above*

Fire set by a farmer races across a field of wheat stubble, north of Wichita, Kansas. Controlled burning to make tillage easier or remove residue quickly for double cropping is short-sighted land management. Burnt-over fields suffer loss of humus, produce lower yields, and are more susceptible to erosion damage.

Deep-rooted prairie vegetation, however, not only survives but thrives under burning, and fire, whether caused by lightning or set by the Indians, helped maintain and extend America's central grasslands before they were converted to agriculture. Traveling in Kansas in 1832, the painter

George Catlin described the chameleon character of prairie wildfires: "Where the grass is thin and short, the fire slowly creeps with a feeble flame, which one can easily step over...These scenes at night become indescribably beautiful, when their flames are seen at many miles distance, creeping over the sides and tops of the bluffs, appearing to be sparkling and brilliant chains of liquid fire...suspended in graceful festoons from the skies." In contrast, there was the "hell of fires! where the grass is seven or eight feet high...and the flames are driven forward by the hurricanes, which often sweep over the vast

prairies...The fire in these...often destroys, on their fleetest horses, parties of Indians...the high grass is filled with wild pea-vines and other impediments, which render it necessary for the rider to guide his horse in the zig-zag paths of the deers and buffaloes, retarding his progress, until he is overtaken by the dense column of smoke that is swept before the fire...kindling up in a moment a thousand new fires...rolling on the earth, with its lightning's glare, and its thunder rumbling as it goes.'

131

● *Top left*

WINDROWS CLEAVE to the curves of steep terrain near Paso Robles, California. The cross-slope cropping pattern of barley helps control erosion, but the farmer's main concern may have been maneuvering equipment on slopes ranging up to 35 percent or more. The usual cropping sequence for dry farming in this region is a grain and summer fallow rotation. Some very steep slopes and gullies – the scruffy brownish patches – are left in native vegetation. The weather in the area is highly unpredictable, with rainfall that varies from as little as eight inches to as much as thirty-five inches annually, mainly during the winter months. The windrowed fields may have been cut for hay because of poor yield, or possibly frost damage, or simply a need for forage. In the lower field of standing grain, there is evidence of lodging.

● *Bottom left*

RED HOT agricultural product, chile peppers are hand harvested by Mexicans who cross the border daily to work in fields near Las Cruces, New Mexico. As the taste for fiery food has grown – not only for Tex-Mex cooking but for Chinese, Indian, African, and other spicy cuisines – so has the acreage planted to chiles, and the New Mexico crop is now valued at about $40 million. Planted in March in irrigated fields in the Rio Grande Valley, chiles are picked green in August and the mature red pods are harvested in September and October. Chile peppers, which originated in Mexico around 7000 BC, range from the relatively mild Anaheim to the incendiary Sandia. Both red and green chiles are used fresh, dried, canned, and frozen in salsas, marinades, barbecue sauces, and other condiments.

● *Top center*

CURRENTS AND eddies in a river of gold mark the trail of a combine harvesting winter wheat near Salt Lake City, Utah. To protect the land from erosion during the coming year, when it will be fallowed, the farmer has left crop residues in place as "stubble mulch." By reducing evaporation and trapping snow, standing stubble also helps the soil store up moisture in a region where the summers are hot and dry and most of the precipitation comes as snowfall in the winter.

● *Top right*

TONGUES OF flame, feeding on dry stubble and whipped by winds, uncurl across a wheatfield in southeastern Washington. Under such conditions, fires, either deliberately set or ignited by lightning, can escape and quickly spread out of control. The toll on farmland is a litany of losses: loss of organic matter in the soil, loss of soil moisture, and loss of the soil itself through erosion due to loss of protective cover.

● *Bottom center*

SPINNING OUT straw and chaff in windrows, a farmer harvesting small grain weaves a beige-on-brown tapestry, near Bozeman, Montana. The residue will probably be baled for cattle feed or bedding. The farmer is working alone, operating the truck as well as the combine, which may account for the varying length of the passes. When the grain tank on the combine fills up and he must auger the grain into the truck, both time and fuel would be wasted if the vehicles happened to be at opposite ends of the field.

● *Bottom right*

TWO COTTON pickers head down the homestretch toward a trailer that will carry their harvest to a gin near Greenville, Mississippi. Mechanization came late to cotton culture. In 1940 hand labor was still used for planting, cultivating, and harvesting, but World War II shrank the labor pool, and by the mid-1950s farmers were using machinery for every task in the crop cycle.

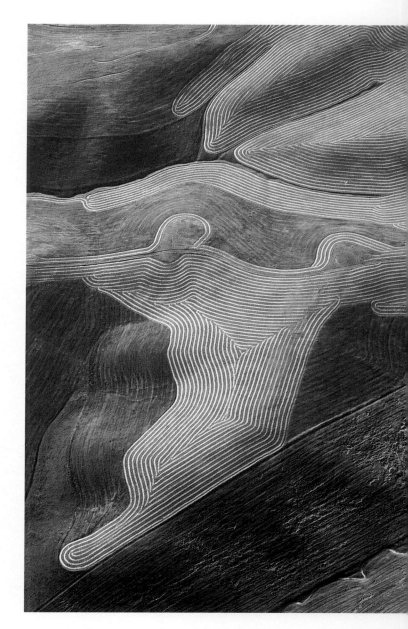

132

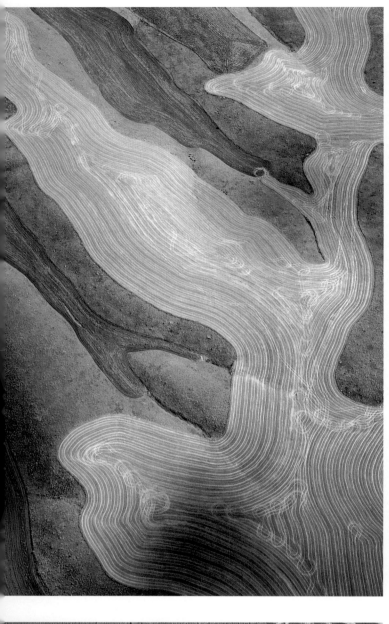

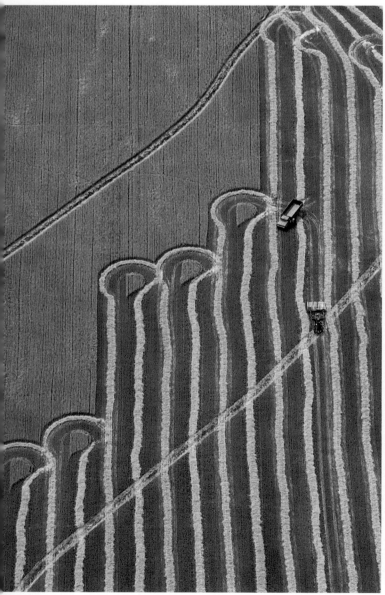

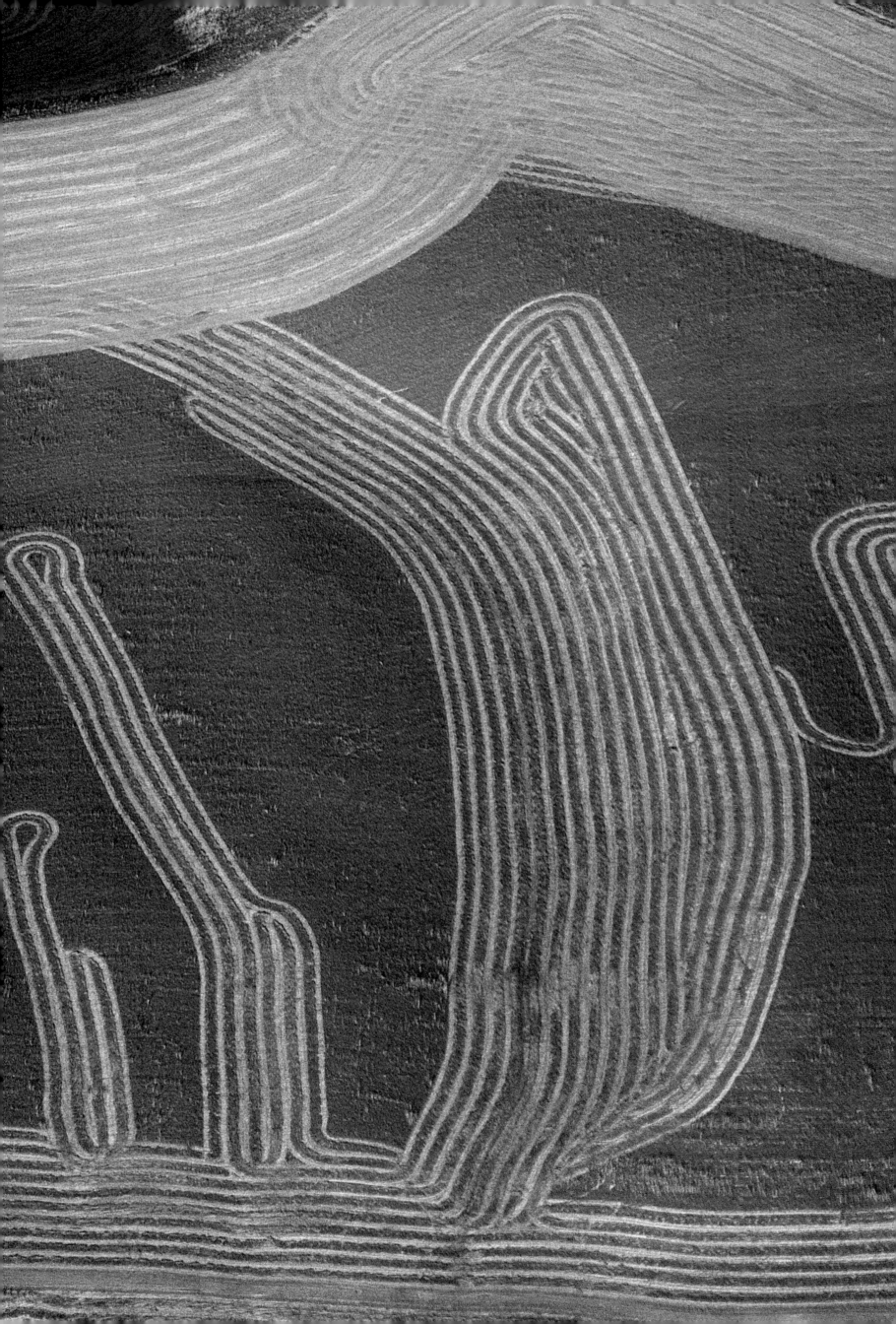

● *Right*

CIRCLING IN on a last stand of grain, harvesters have gathered in most of the yield from a field near Kremlin in Montana's Golden Triangle, an area noted for production of three small-grain cereal crops: winter wheat, spring wheat, and barley. Combines have spread wide trails of straw and chaff behind them, which are overlaid with the narrow tracks of trucks coming and going as they collected the grain from the combines. The island yet to be harvested may be a low spot where snowmelt or spring rain ponded, causing the grain to put forth vigorous growth but to mature at a later date.

● *Right*

RECTANGLES BECOME diamonds in a giant cat's cradle created by barley harvesters near Fairfield, Montana. After the grain was swathed for even ripening, a combine with a pickup head gathered up the grain and threshed it, and left the straw in windrows to be baled for livestock feed or bedding.

● *Previous pages*

DESIGNS DRAWN by a farmer in a lentil field in eastern Washington call to mind the enigmatic lines etched by the ancient Nazcas on a desert plateau in southern Peru. Higher soil moisture and/or lower soil temperature apparently slowed ripening of the lentils in some areas, so to hasten the process the farmer has cut the crop and laid it in windrows with a swather. Later he will complete the harvest with a combine, as he already has on the upper part of the slope.

● Left

CARVING A crop of hay out of a carpet of coastal Bermudagrass, near Trenton, South Carolina, two machines work in opposite directions but to the same end. The golden strips in the center are hay that was mowed earlier and left to air-dry in the sun, since too much moisture can cause baled hay to rot or even catch on fire from spontaneous combustion. A tractor-drawn hay rake is now windrowing the hay into narrower strips for easier handling by the follow-up baler. Traveling at a much slower speed, the tractor-drawn hay baler is just beginning its first circuit, rolling the hay into big round bales.

● Left

KENTUCKY BLUEGRASS has gone to seed – literally but not figuratively – in ponderosa pine country, near Reubens in western Idaho. A perennial grass used for turf and pasture, the crop seen here has been grown for seed and is now being harvested. Around the edges of the fields and in a broad central band, the grass has been swathed to accelerate ripening, and in some of these areas a combine has already scooped up the windrows and separated the seed from the straw. Between the two fields, a truck appears to be fitted with bulk racks to transport the seed to the warehouse. In the unharvested portions the grass has lodged in some places, probably because of a rainstorm when the seed heads were full and heavy. If the lodging occurred early in the growth stage, the stems could be broken and the crop could be lost, but if it occurred after the seeds had formed they could continue to mature and produce a good yield.

● *Above*

COMBINES HARVESTING rice cut straight across meandering levees near Stuttgart, Arkansas; before the fields are replanted, the levees must be rebuilt. During the growing season, the levees hold water on the fields for irrigation and weed suppression, but gates in the levees are opened before harvest, allowing the water to run off and the land to dry, so heavy equipment does not bog down. In most of the world rice cultivation remains labor-intensive, but in the United States rice production has become highly mechanized.

●

Rice combines are specially designed for operation on soft, muddy ground, as are rice dollies which haul rice from combines to trucks. Other equipment used by rice growers includes field cultivators and disks to loosen the soil and land planes to smooth it, laser-equipped tractors to fix levees on the contour and levee plows to construct and maintain them, and tractor-drawn grain drills to plant the rice. In some areas, airplanes are used to plant rice, and most fertilizers and pesticides are applied aerially.

● *Right*

HARVEST HOME! Spewing out residue behind them, one combine heads for the bulk truck as another veers back to harvest the last slim wedge of peas in the field in eastern Washington. But harvest – good or bad – brings no rest for the farmer. Time, the weather, and the agricultural cycle wait on no one, and as soon as the harvest is in, the nation's farmers must begin preparations for planting next year's crop for, when the sun comes up tomorrow, once again the world will wake up hungry.

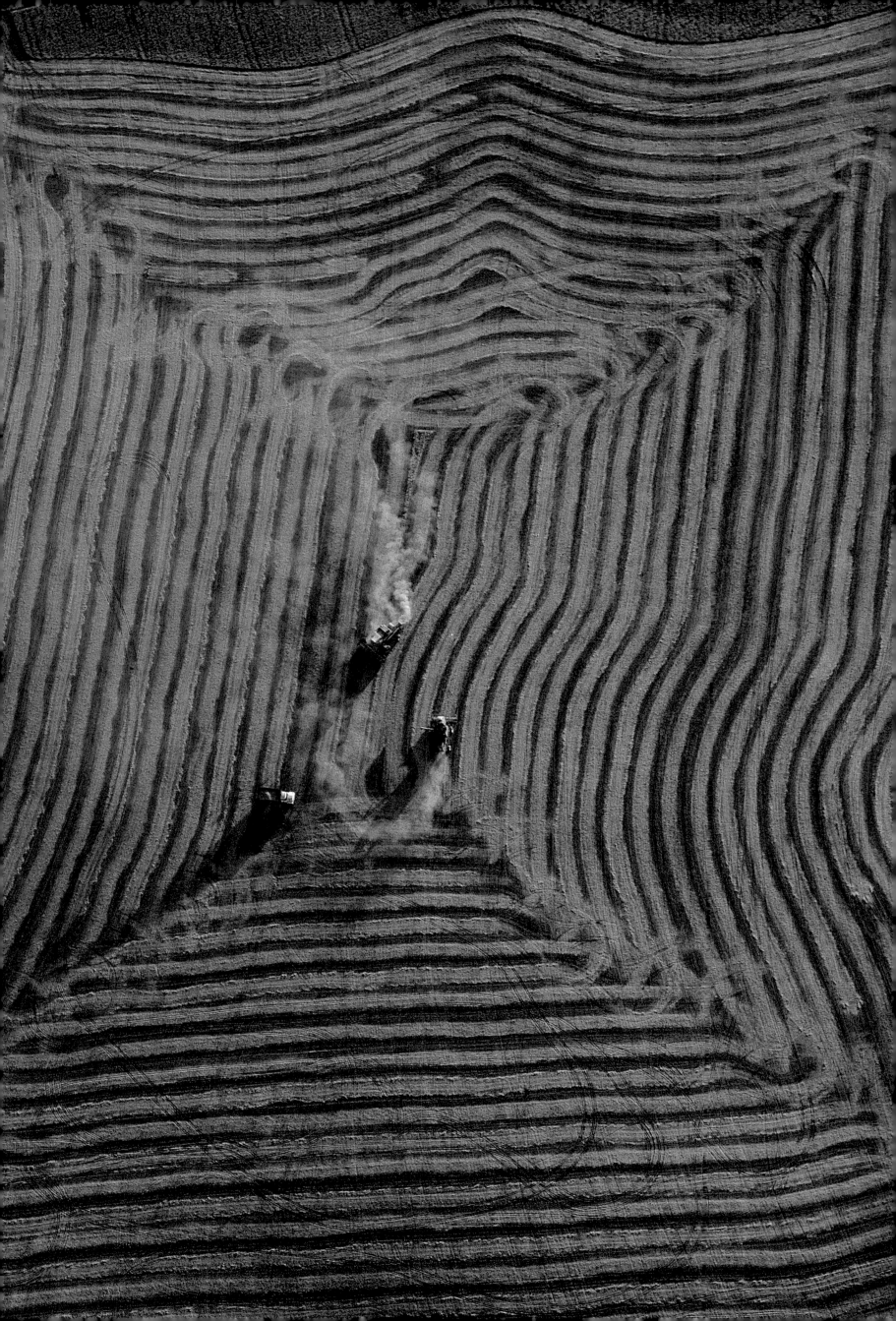

COPING WITH THE ELEMENTS

"In the lotteries of human life . . . even farming is but gambling," Thomas Jefferson concluded in 1813. Today, farmers' fortunes still depend on nature's roll of the dice. As unpredictable as the weather may seem, it's a sure bet it will never be "normal." We can be certain the benign rhythm of rain and shine will be punctuated by droughts and blizzards, floods and tornadoes, withering heat waves and killer cold snaps – the question is when? Farmers can choose crops that are early maturing or drought resistant. They can supplement scanty rainfall with irrigation, and shelter crops from wind and frost. But farmers cannot control the weather, and the age-old cycle of sowing and reaping still follows the seasons. Farmers place their bets at planting time, hedge them during the growing period as best they can, and pray the payoff will be a bin-busting harvest.

Nowhere are the hazards greater or the stakes higher than in the nation's midsection, where the weather can deliver bumper crops or a punishing blow to the solar plexus. A few soaking spring rains are needed to get the corn crop off to a strong start. Some years, however, fields are too soggy to plow. When warm, humid air from the Gulf of Mexico meets cold, dry air spilling over the Rockies, cumulonimbus clouds boil up, generating some of the most destructive weather on earth: Thunderstorms trigger floods that wipe out crops and sweep away livestock; lightning strikes farmers in their fields and sets hay barns ablaze; raining down like buckshot, hailstones riddle fruit and shatter greenhouses; and twisters ricocheting up Tornado Alley splinter storage bins and toss tractors end-over-end.

As spring turns to summer the specter of drought haunts the Great Plains. The region was once known as the Great American Desert, and farming remains chancy. In 1988 parched crops shriveled and died, soil cracked and crumbled, as a cantankerous combination of heat and drought caused a $10-billion crop failure, the costliest in American history. Fall rains benefit newly planted winter wheat, but heavy downpours can damage corn almost ready for harvest. Too soon, the brutal cold of winter sets in. Winter wheat can withstand subzero temperatures if blanketed by snow, but it may succumb when winds out of the Arctic send the mercury tumbling to thirty below. Cattle on the range can weather most blizzards by drifting before the storm, rumps to the wind. Other livestock are not as hardy, and during severe winters early settlers often shared their sod houses with horses, pigs, and poultry, and even shared their beds with seed potatoes to keep them from freezing.

Among other natural phenomena affecting farmers, most important is erosion. The work of wind and water, erosion wears away rock and creates new soils along with other geomorphic processes. America's most fertile farmlands have evolved from volcanic deposits, glacial drift, wind-blown loess, and water-borne alluvium. But even as they build, soils, too, are worn away in an endless cycle. Today, human activity has accelerated erosion, and soil that took centuries to form can wash or blow away in a season. Water erosion predominates in the humid Southeast, the hilly Northwest, and the Corn Belt, where Iowa has lost half of its topsoil in scarcely a century of farming. Wind erosion is worst on the western flatlands; the winds above the Mississippi basin have been estimated to have 1,000 times the soil-carrying capacity of the river itself.

Few prospects are more menacing for Great Plains farmers than the approach of the wind erosion season when their fields are dry and in condition to blow. Memories of the Dust Bowl run deep. But it was not entirely a natural disaster: Conspiring with heat and drought, decades of often heedless agricultural expansion culminated in the Dirty Thirties. Yet too many farmers continue to exploit erosion-prone land. Abandoning protective cropping patterns, they put all their chips on cotton, corn, or soybeans, notorious "soil skinners." Insensitive to the contours of the land and the will of the wind, they drive ten-plow tractors in arrow-straight runs over treeless terrain. It is within our power to cut our losses of farmland to erosion; nature does not hold all the cards. But the extent to which we do so depends on how we play our hand.

Overleaf: Runoff claws gashes in the red soil of Oklahoma, despite a wheat farmer's efforts to control erosion through diversion terraces in the left field and a new waterway, still lacking grass cover, in the right field. Nationwide, water causes 65 percent of farmland erosion. In 1977, for every pound of food consumed, America lost twenty-two pounds of soil to runoff.

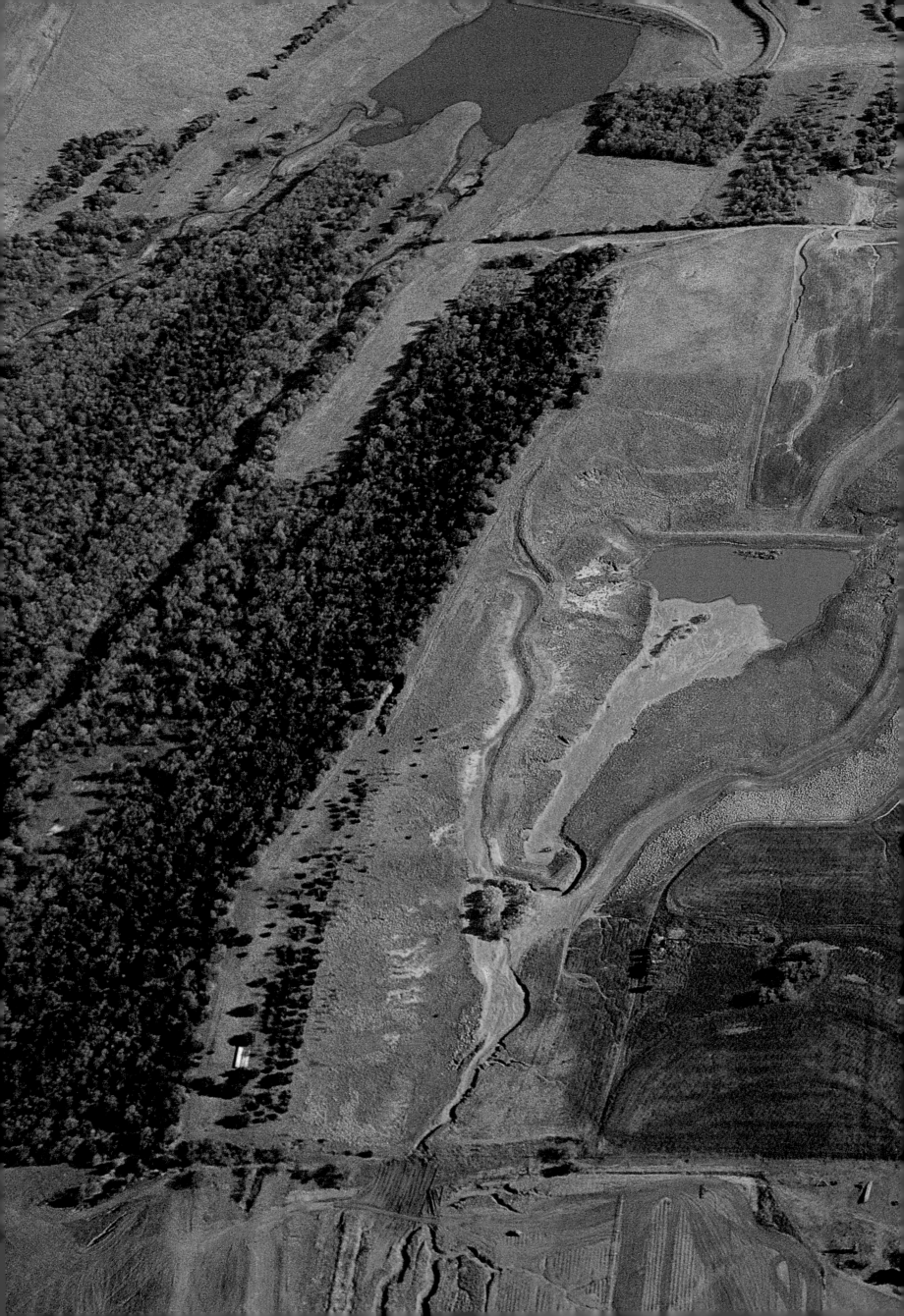

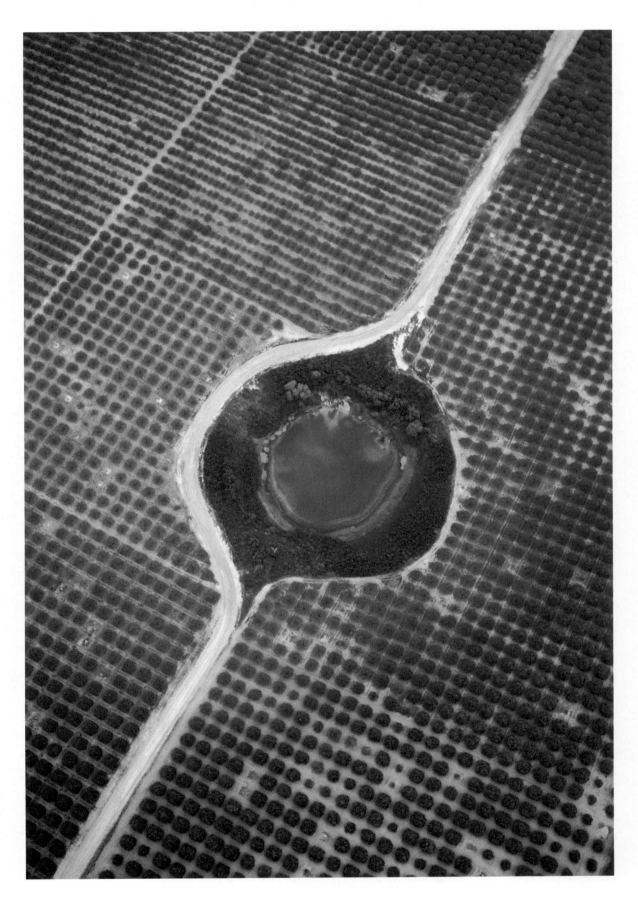

● *Above*

FROSTBITTEN CITRUS trees cluster near
a pond in Orange County, Florida,
where freezes in the mid-1980s
destroyed 80 percent of the trees.
To insulate trees against killing
cold, growers use sprinklers to
coat them with ice.

● *Right*

WHEN WINDS threaten to sandblast
cotton seedlings, Texas Panhandle
farmers roughen their fields with
60-foot-wide "sandfighters" to hold
the soil in place. A lime-ringed
playa testifies to wind erosion in
ages past.

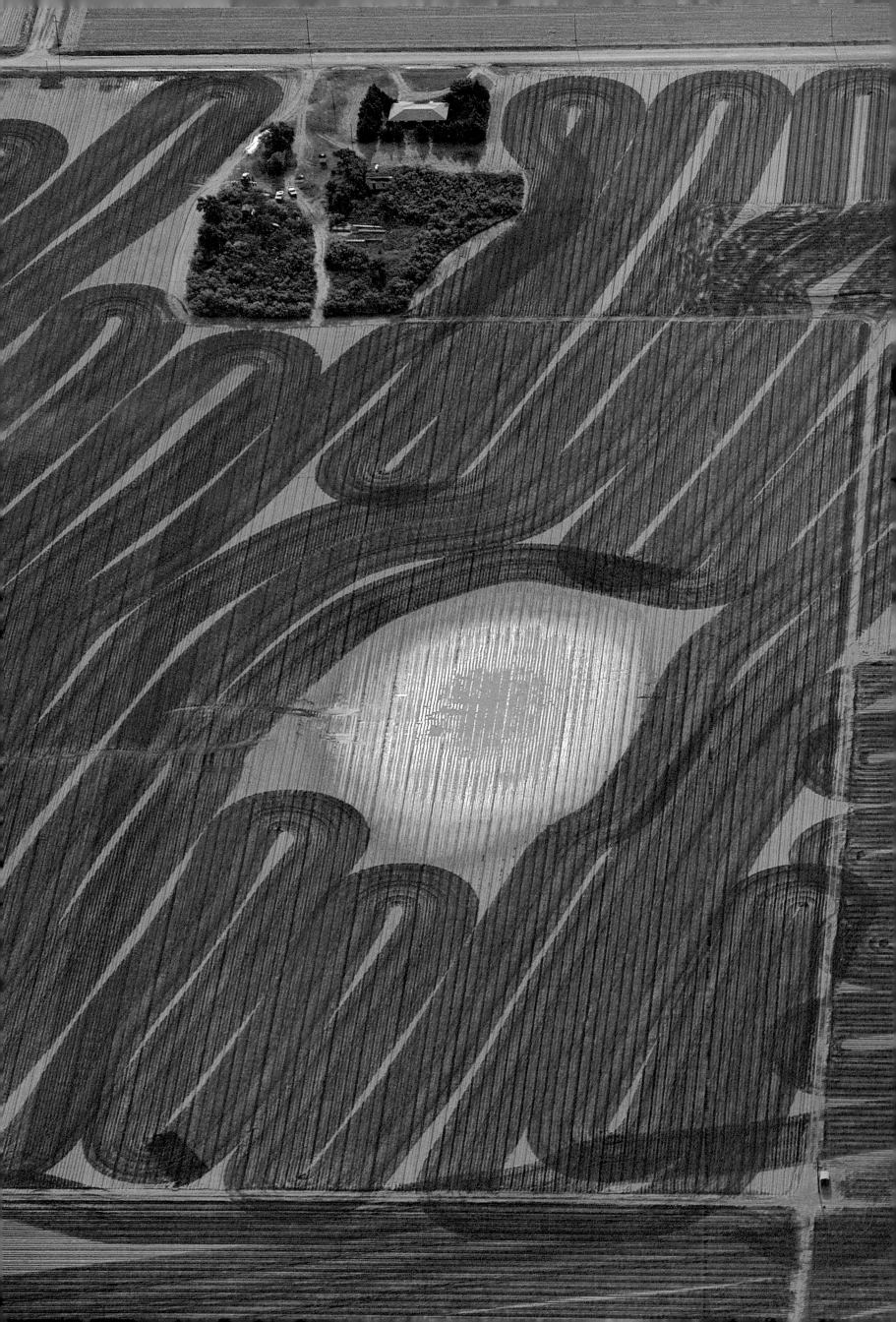

● *Right*

ICY SWIRLS of contour stripcropping joined by frozen seams of grassed waterways reveal topographical nuances in Minnesota, where winter brings four months of snow cover plus a fifth month of chill weather. Alternate freezing and thawing can break up topsoil and lead to erosion unless fields are protected by snow or crop residues. By putting off tillage from fall until spring, farmers can cut soil losses substantially, but when winter lingers on too long, they may fall behind in plowing and planting. Then crops risk damage from early frost before they are ripe and ready for harvest.

● *Right*

DUSTING OF snow enhances imprints left by fall tillage in Minnesota. The likely scenario: When field "trash" clogged a farmer's plow, he would make a circular swing out of the furrow, raise his plow and shake it free of debris, then resume plowing. Despite considerable snowfall, little infiltration occurs when the ground is frozen solid, and strong winds blow snow off bare fields. The fast-spreading practice of reduced or "conservation" tillage, however, leaves standing stubble or surface residues that not only hold the soil but trap snow. The subsequent absorption of meltwater permits annual cropping in rain-short areas where formerly crops were grown only in alternate years and fields were left fallow between crops in order to accumulate moisture.

● *Left*

SCUDDING ACROSS a snow-swept landscape, a square-rigger of manure bears a cargo of nutrients for next summer's crops. When fields are not buried in snow, Minnesota farmers apply both liquid and solid livestock wastes without plowing them under. Manure spread on frozen ground or snow, however, risks being flushed away by rapid spring thawing and discharged into surface waters. A 1984 survey found agricultural runoff to be a major contributor to the pollution plaguing more than 75 percent of Minnesota's lakes.

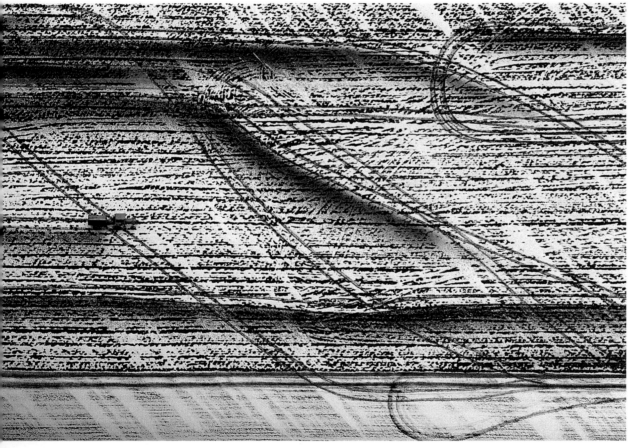

● *Left*

PERFORMING A cold and lonely task, a Minnesota farmer drives a tractor trailing a manure spreader, which unloads its burden off the side. Dark streaks mark earlier runs. Unless a farm is equipped to stack manure, barnyard waste must be disposed of by applying it to fields. Frigid temperatures, biting winds, and heavy snowfall, which can average eight feet or more in northeastern counties, make hard farm chores even harder: cleaning out snow-bound barns, chopping ice off watering troughs, and feeding livestock – whose need for feed rises as the mercury falls – from mid-October until the end of April. Cold takes its toll, and spring blizzards can kill newborn calves. One consolation: subzero temperatures can also kill insects, thereby reducing damage to next season's crops.

● *Above*

BATTERED BY heavy spring rains and high winds. lodged grain forms a tangled mat in Tidewater Virginia. Bent but not broken, the young wheat or barley will recover, but lodging later in the season could cause serious losses.

● *Right*

DROUGHT EATS a hole in a carpet of sunflowers in North Dakota. Sucking up subsoil moisture with long taproots, sunflowers can weather most dry spells, but not the searing summer of 1988, which caused record crop losses.

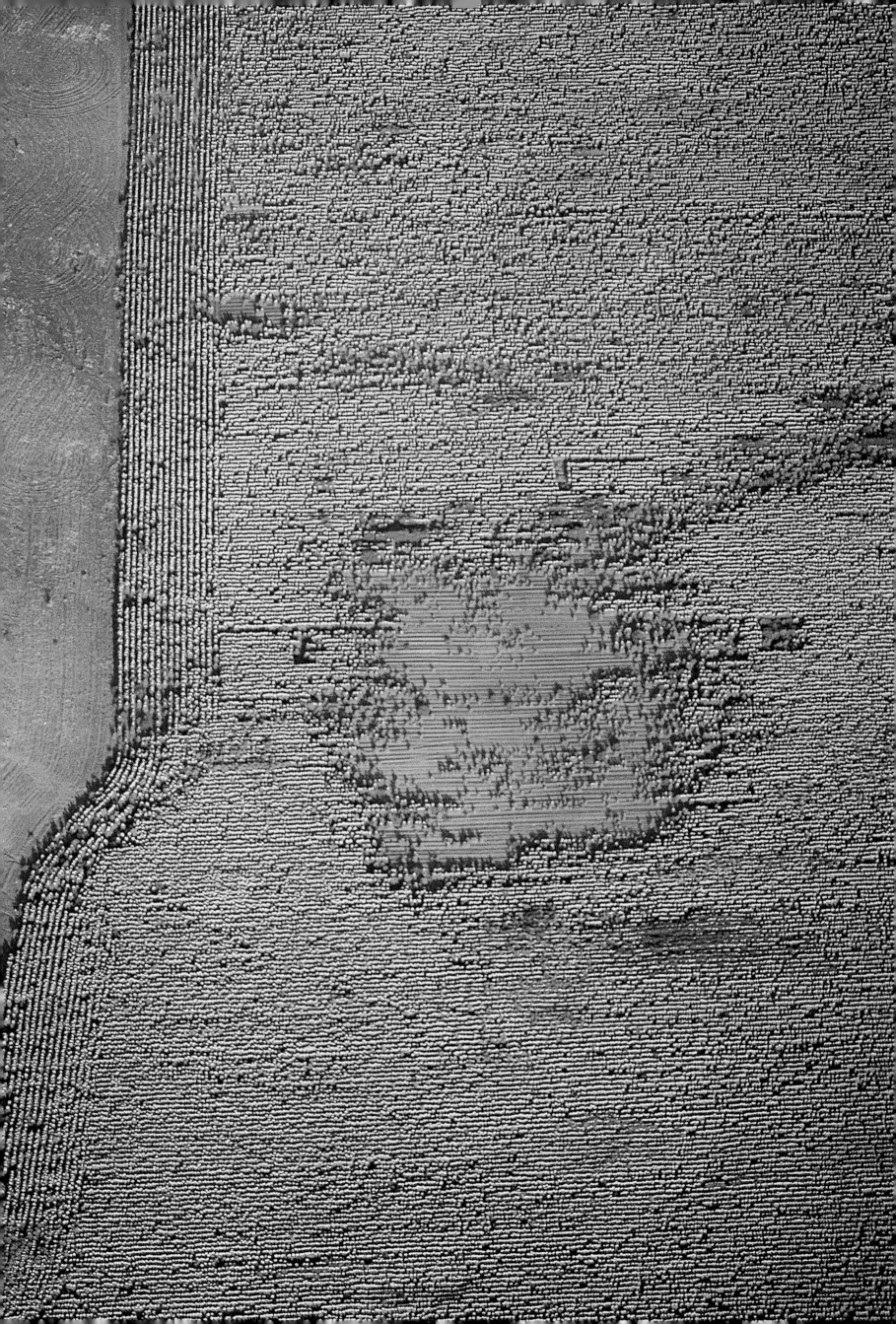

● *Above*

"SOILS THAT melt like sugar and flow like water" make for chronic erosion in the South, as in this peanut field on the coastal plain of Alabama, where ephemeral gullies scar land depleted earlier by continuous cotton cropping. Erosion is most ruinous in the spring, when newly sprouted row crops afford scant protection against violent rainstorms. According to a 1982 survey, on 65 of every 100 acres planted to row crops in Alabama, soil was being lost at greater than tolerable rates – that is, faster than it is generated.

● *Facing page*

"DROWNED OUT," green fields of burgeoning winter wheat will never mature for harvest. "Washed out," brown fields of grain sorghum planted in May must be resown in June; a farmer has disked around ponded areas to hasten drying. A surfeit of rain – an infrequent event in central Kansas – creates havoc for Plains farmers. Against the threat of wind erosion, however, they have built strong defenses: a shelterbelt of trees reinforces alternating strips of field crops laid out at right angles to the prevailing wind.

● *Overleaf*

GHOST IMAGES on the South Plains of Texas reveal the sluggish curve of a filled-in drainageway and the shadowy rim of an ancient playa, or desert basin, formed by wind and water over geologic time. Today the cultivation of cotton, the prime money crop of the region, accelerates ongoing erosion by leaving the loose soils exposed to high winds during the November-to-May blow season. Recurring drought exacerbates the problem. Growing forage sorghum – on set-aside land at left center – is a good conservation practice but no way to make a living.

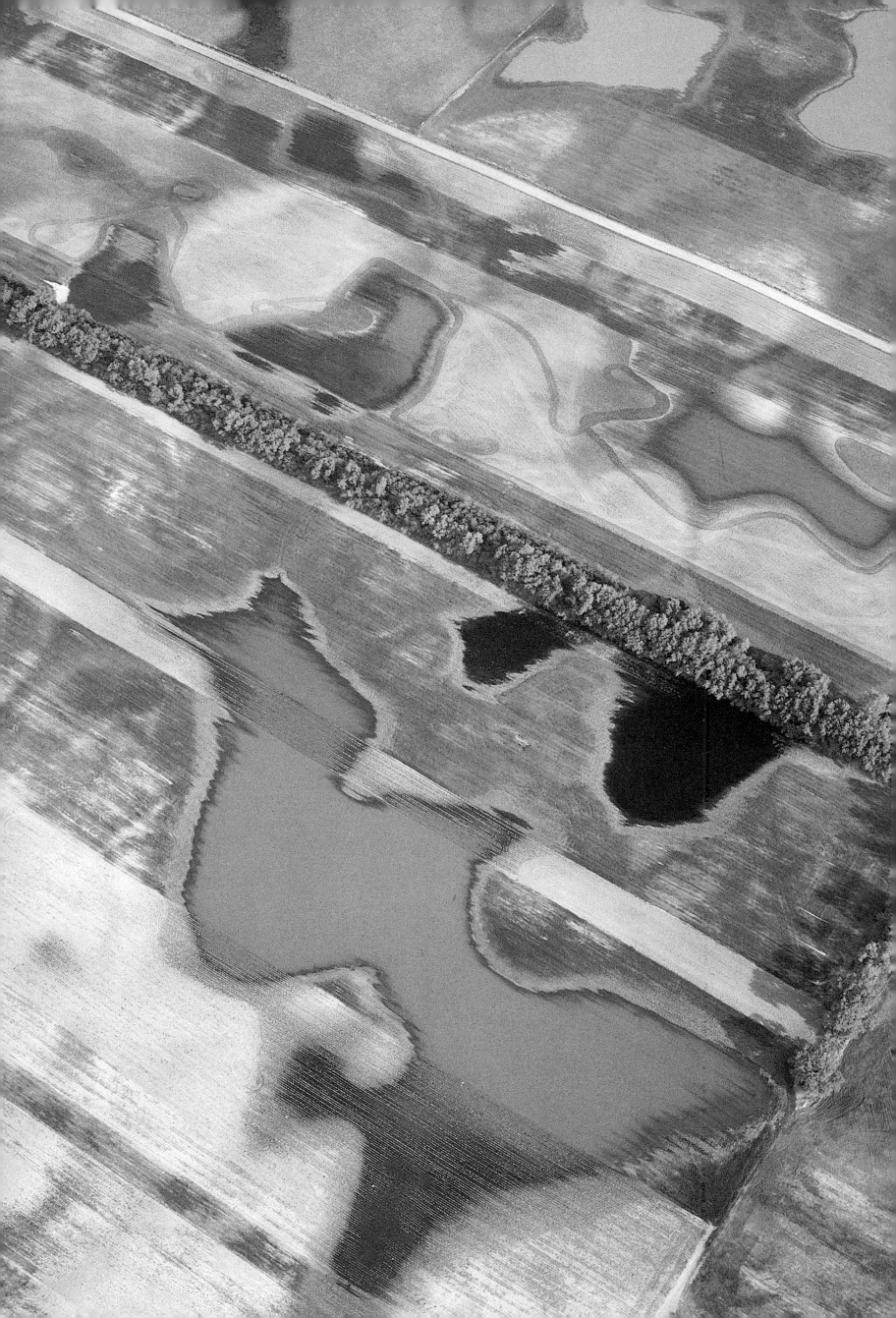

● *Above*

UPROOTED TREES and farmland
buried in sand and silt emerge
from the wake of the rampaging
Cimarron River in Oklahoma. In a
costly salvage effort, a landowner
bulldozes sand to fill in a pothole
scoured out by the flood.

● *Right*

LOBLOLLY PINES spring from a ravine
carved by gully erosion in the
loamy soils of Alabama. On-site,
erosion reduces cropland productivity; off-site, the soil that is washed
away clogs and contaminates
streams and lakes.

● *Above*

SEDIMENT BASINS, built astride runoff paths, trap soil eroding from a soybean field in Tennessee. However, nearby Reelfoot Lake, formed by an earthquake in 1811, is doomed to extinction in 2032 at present rates of sedimentation.

● *Right*

LIKE DRIFTED snow, volcanic ash blown aloft by Mount St. Helens in May 1980 settled in lentil and pea fields surrounding a stand of barley in eastern Washington. In some areas the ashfall was so heavy it damaged fruit trees.

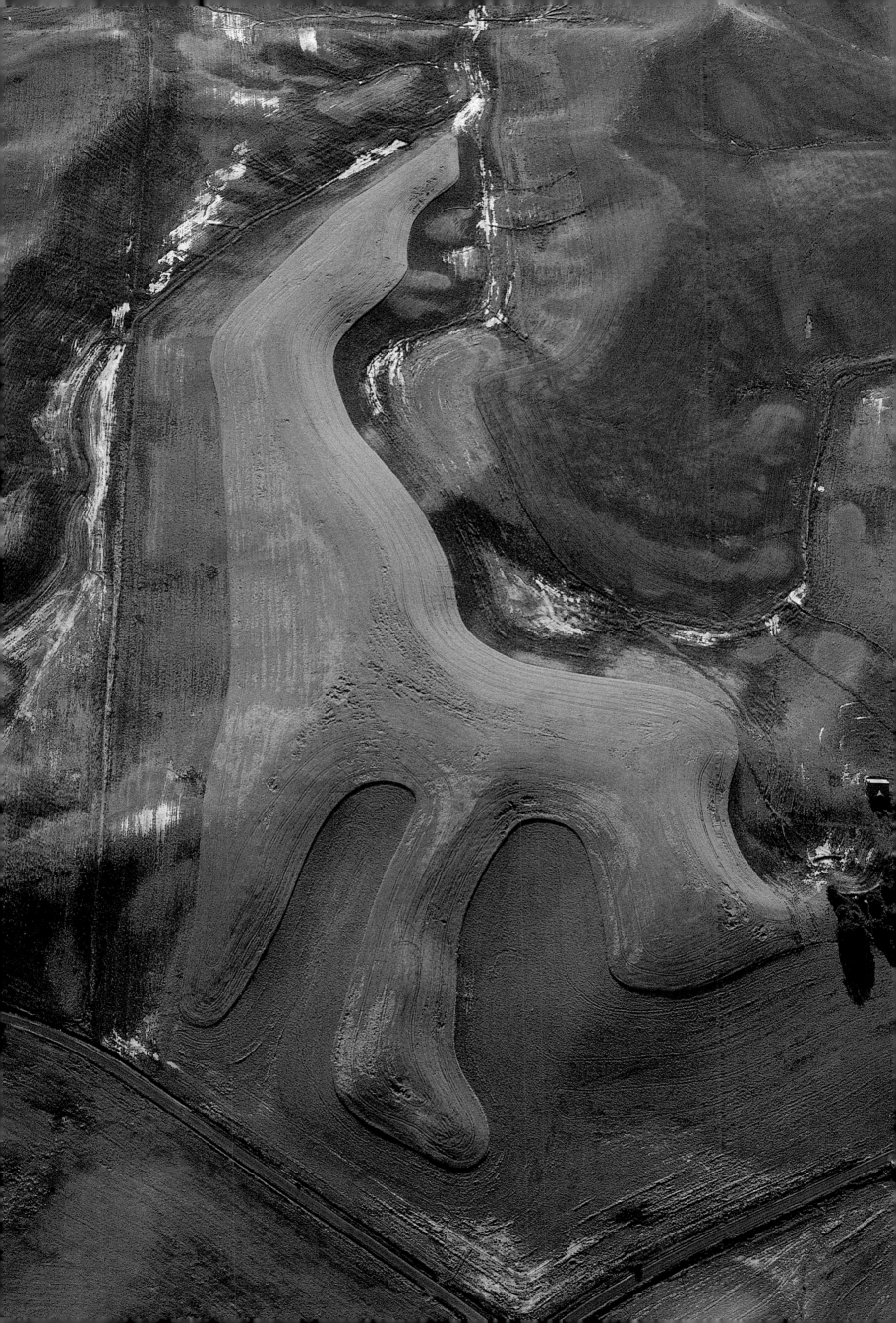

BATTLING EROSION

"**F**arm ugly!" The latest rallying cry against erosion exhorts farmers to leave their fields untilled and littered with residues – called "trash farming" in some circles. The erosion of farmland, and how to stanch it, has always stirred passions. Appealing to the civic zeal of fledgling Americans, Patrick Henry reportedly declared, "He is the greatest patriot who stops the most gullies." Others have seen good land stewardship as a religious duty. Like Adam, who was put in God's garden "to dress it and to keep it,' the Amish feel bound by divine injunction to cherish the corner of creation entrusted to them. Yet from colonial days, when fields "wore out," farmers moved West and broke new ground. The plow that broke the Plains, however, also exposed them to the elements. Too often, fields dried up and blew away, until the dust storms of the 1930s shocked the nation into mounting an offensive against erosion.

A single raindrop can crater soil like a tiny bombshell; a cloudburst can loose a barrage of 200 tons of water per acre. A breeze can set soil particles bouncing; gale-force winds can whip up gritty blizzards. One of the oldest strategies to combat erosion is to plow along the land's contours instead of up-and-down hill. The level furrows slow runoff, reducing soil losses by as much as half. Extolling the virtues of contour farming, Thomas Jefferson claimed, "In point of beauty nothing can exceed that of the waving lines and rows winding along the face of the hills and valleys." Today the fastest-growing erosion-control practice is conservation tillage, in which farmers plow less or not at all, leaving residues on the surface to shield the soil from rain and wind. No-till may look slovenly, but it can cut erosion by 95 percent. Using crop rotation, growers plant in sequence row crops such as corn or soybeans, which provide scant cover and drawn down nutrients, followed by sod-forming grasses or certain legumes, which produce a thick canopy and build up the soil. With strip cropping, farmers create vegetative barriers to wind and water erosion by alternating bands of primary crops with dense stands of grass or alfalfa. Shelterbelts of trees and shrubs form permanent windbreaks, while emergency tillage, or "sandfighting," roughens the soil to increase wind resistance. Terraces intercept runoff, allowing water to be absorbed and soil redeposited, and excess water channeled off slowly. Grassed waterways conduct runoff safely down a slope.

Yet despite an extensive arsenal, half a century after declaring war on erosion, the United States was losing topsoil faster than at any time in history – more than three billion tons annually. Hadn't we solved the problem? We seem to have suffered no decline in productivity. Nor loss of cropland: Cultivated acreage rose to an all-time high in the 1980s. But increased yields since the 1950s have been due largely to technological advances, which tend to mask the effects of erosion. And policies that urged farmers to plant "fencerow to fencerow" to meet world demand in the 1970s also led them to plow under grassed waterways, bulldoze terraces, and abandon other conservation practices that fast, wide-swath machinery had made "inefficient." Farmers know how to control erosion, but whether or not they do so is usually dictated more by dollars-and-cents economics than by ethics or emotions.

In landmark legislation, the Food Security Act of 1985 recognized that the nation's soil resources are the heritage of all Americans, and that the costs of conservation should be shared through government funding. The Conservation Reserve Program offered farmers strong incentives to retire some 40 million acres of highly erodible cropland from cultivation, with the government paying rental and defraying half of the costs of planting the land to grass, trees, or wildlife cover. Farmers who continue to crop highly erodible land must install an approved conservation system or lose their eligibility for Department of Agriculture benefits. By 1989 these measures had already reduced soil losses by about a third and could cut them in half by 1995. That would be half the battle. Throughout history, farmers have changed their patterns of production to meet new challenges. The need now is to develop ways of farming that permit use of the land without loss of the soil, patterns for an enduring agriculture.

Left: Striped mantle of wheat and fallow conserves soil and water in western Nebraska. Cropped strips form barriers against wind erosion, while fallow strips store up moisture for next year's crop.

Overleaf: Mosaic of erosion control practices marks dairy country near Fond du Lac, Wisconsin. Crops for feed are rotated on strips that are straight or contoured, depending on the terrain. Other features of good soil management include grassed waterways, woodlots, and wildlife habitat.

● *Left*

HOLDING THE red soil in check, contoured terraces provide a foothold for wheat in north central Oklahoma. The wheat on the right will be harvested for grain; that on the left, where cattle are grazing, serves as livestock forage. Breaking a slope into a series of steps — the steeper the slope, the narrower the steps — terraces control erosion by slowing runoff and its suspended sediment, then channeling excess water off the fields. Usually highly efficient, terracing can reduce water runoff by as much as 90 percent, and can cut soil losses by

75 to 95 percent in most cases. But torrential rains may overwhelm a system, and the violent caprice of midwestern weather probably caused the gully erosion seen at the ends of the terraces. Although initial construction of these earthen embankments is expensive, terraces can remain effective for years if properly maintained. Many have been torn out, however, because they are difficult to farm with the wide, heavy equipment now commonly used by American agriculture.

● *Above*

TARGETING SOIL erosion, the bullseye was laid out during "Strip Week" in the spring of 1988 in Fauquier County, Virginia. Sponsored by the local Soil Conservation Service, the event promotes stripcropping as a way to preserve soil resources. The circular pattern, dictated by the terrain, controls soil erosion on a knoll that slopes down from the center uniformly in all directions. The green crop is a pasture mix of grasses and legumes; the bare strips will be planted in corn. The gully at right will become a grassed waterway, providing a safe outlet for runoff.

● *Top left*

A COLLAR of ripening corn circles a
hillock of hay near Canton, Ohio.
The farmer will husband the soil
by regularly rotating the crops
and reversing the pattern. In flat
regions of the Corn Belt, fields
are sown mainly to corn, with
soybeans an important second
crop; but in hilly areas, which are
subject to water erosion, these row
crops are alternated with sod-
forming forage crops such as
alfalfa, clover, or meadow grasses.
Corn is soil depleting: it offers little
protection against erosion and
draws heavily on soil nutrients.
Grasses and legumes, however,
are soil conserving: they provide
a thick protective cover against
erosion and return organic matter
and nitrogen to the soil. In part,
the strips and circles follow the
contours of the land – another
conservation practice.

● *Bottom left*

PRIZE-WINNING PATTERNS of
husbanding the soil as well as beef
cattle earned Don Dowler the title
of "Conservation Farmer" for
Marshall County, West Virginia, in
1988. His exemplary practices
include plowing and planting on
the contour and cropping alternate
strips of dark green alfalfa hay and,
on the beige strips, corn, which
has been cut for silage. The lighter
green island is reserved for
pasture, as is the surrounding area,
with grass on the benches and
bands of oak and locust trees on
the steeper parts of the topography.
In this region of precipitous
hillsides and narrow valleys, farms
are often located on the ridges.
In the spring the soil dries out
and warms up sooner, permitting
earlier planting; with good
exposure to the sun, hay cures
faster; and there is little risk of
flood damage.

● *Top center*

"SEATBELTS FOR the soil," one
conservationist calls them. Contour
strips planted to corn, small grain
and hay hold the earth in place on
the rolling hills of northeastern
Ohio. Contour farming involves
plowing, planting and harvesting
along the contours of the land.
The Soil Conservation Service
assists in the design of contour
systems and stakes the lines in the
fields, but fewer farmers now
choose this method of tillage
because they find it uneconomical.

● *Top right*

CURVING STRIPS enfold a mountain
meadow on the rugged Central
Allegheny Plateau, near
Morgantown, West Virginia, where
contour farming is essential to
combat erosion. Corn grown on
the strips and the hilltop oval has
been harvested and stored in the
silo for winter feed for dairy cattle.
Also on the menu will be hay cut
from the meadow, which has been
rolled into large round bales
ranged along the road.

● *Bottom center*

SPRING GREEN of sprouting oats sets
off the warm earth tones of fields
freshly planted to corn, or ready for
planting, near Ripon, Wisconsin.
The crops will provide hay and
silage for dairy operations. In a
field near the farmstead, cows are
grazing. Contour stripcropping
helps protect the sloping terrain
against erosion. In addition,
grassed waterways conduct
excess runoff from the fields
without damage.

● *Bottom right*

SCULPTURED STAFF of burgeoning
corn and tonsured grass and
legumes echoes the gentle
undulations of the topography near
Canton, Ohio. Erosion control is
built into the patterns of Ohio
farming, bringing stability to the
fertile soils – with rates of erosion
well below the national average –
and a serenity to the landscape
that affirms Aldo Leopold's
premise, "Conservation is a state of
harmony between men and land."

164

● *Above*

SCRAWLING A crosswork pattern on fields "ready to blow," a farmer near Chickasha, Oklahoma, fights to save the sandy topsoil. Normally these fields would be shielded from wind erosion by winter wheat, but either the fields weren't sown in the fall or the crop has failed. When wind threatens, the farmer resorts to emergency tillage, using a chisel plow to bring up clods without turning the earth. The roughened surface withstands the force of the wind and reduces its speed. Farmers usually work strips at right angles to the prevailing wind. This farmer is prepared for an attack from any quarter.

● *Facing page*

PULLING A sandfighter over a cottonfield near Lubbock, Texas, a farmer races a rising spring wind in the battle against erosion. The implement roughens swaths up to 80 feet wide, turning up moist clods along the rows of emerging cotton. Only temporarily effective, the procedure must often be repeated. Texas accounts for one-fifth of the soil losses on U.S. cropland. Weather, soils, and cropping patterns all contribute to the high erosion rate, but it is now declining thanks to alternative conservation methods – windstrips, residue management, and changes in land use.

● *Overleaf*

A GIANT jigsaw of corn and alfalfa secures fertile soils in northeastern Iowa. The interlocking strips follow the land's tortuous contours; cutting across them, grassed waterways carry runoff down the slopes. Once a deep gully, the central watercourse has been planted and stabilized. Compared to straight-row tillage, this complex conservation system reduces erosion by 75 percent. The father and three sons who manage Rolwes Farms exemplify good soil stewardship, preserving the land for future generations.

● *Above*

SHAPING THE land to control erosion regenerates pastureland in southwestern Oklahoma. Two deep gullies that merged into one, downslope, have been filled and smoothed over. Around the top of the slope, a diversion terrace has been constructed to intercept runoff from above and prevent it from entering the erosion-prone trough. The channel that diverts the water will probably empty into a grassed waterway or a pipe outlet; vegetation in the channel will filter out sediments. Such structures can be costly to install but are very effective at reducing erosion and sediment loss. For restoring degraded land, grass, which stablilizes and rebuilds the soil, is unequalled. The small tracks indicate that a Bermudagrass sprigger has been applying "earth's healing bandage," mechanically setting out grass sprigs in narrow rows. A creeping perennial, Bermudagrass will help to hold both soil and water on the land, and provide excellent forage. Controlling runoff and revegetating steeply sloping terrain, such as this, can cut erosion losses from several hundred tons of soil per acre each year to a tolerable rate of only two to five tons.

● *Right*

GRASSED WATERWAYS and looping terraces in eastern Nebraska evoke the rivulets of thawing earth at Thoreau's Walden Pond, taking the form of "leaves or vines . . . resembling, as you look down on them, the lacinated, lobed, and imbricated thalluses of some lichens; or you are reminded of coral, of leopards' paws or birds' feet." Greening the spring landscape, winter wheat, or possibly oats, alternates with ground ready for planting to corn or soybeans. Soil losses could exceed 70 tons per acre per year on these fields, but contoured terraces, grassed waterways, and two erosion control dams keep losses to a minimum.

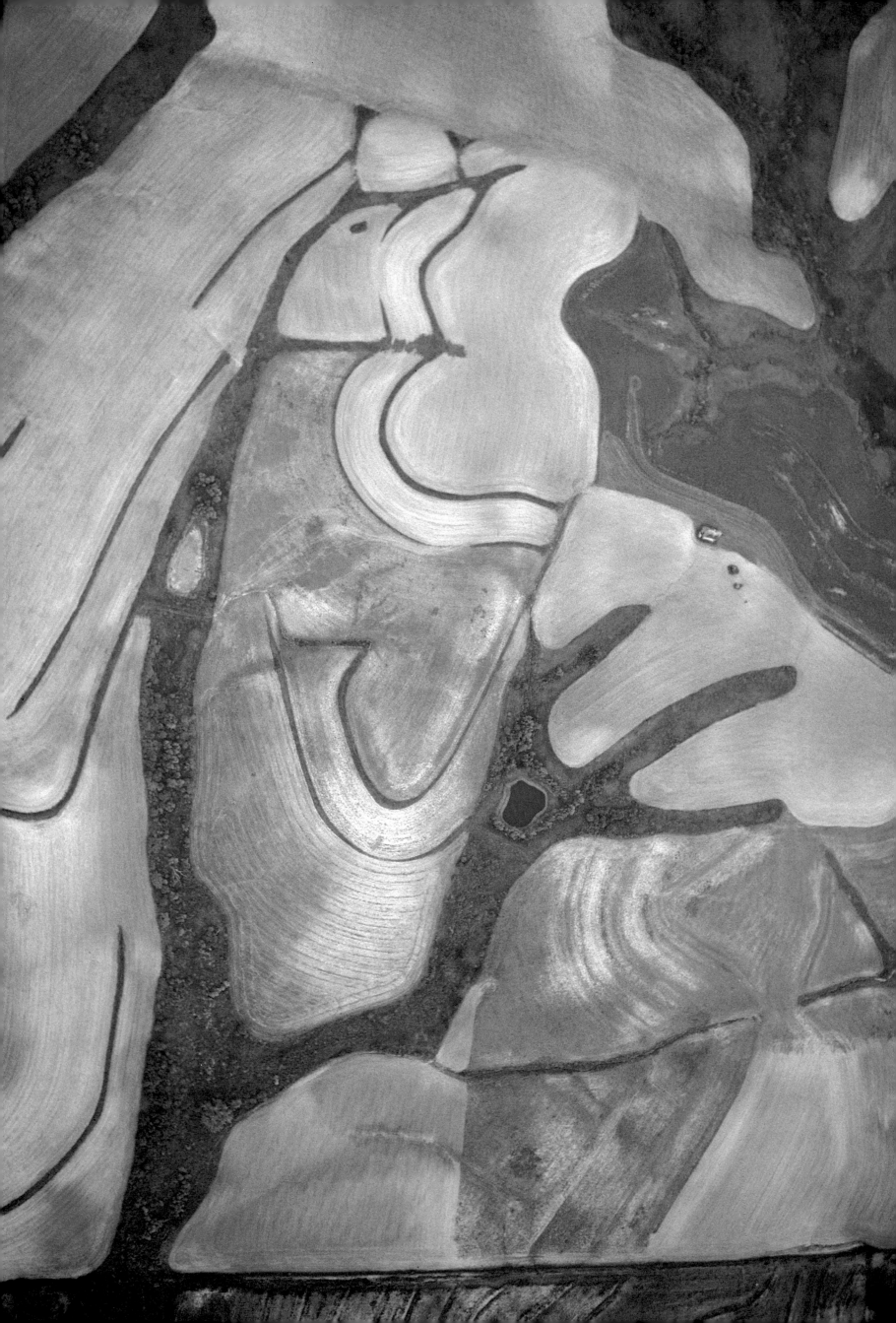

SLAKING AGRICULTURE'S THIRST

"By our faith ... our fields are watered," say the Hopi Indians of the arid Southwest. Reinforcing their faith, the Hopis' intimate knowledge of their mesa world has enabled them to prosper for more than a millennium in a land of sparse rainfall and scarce surface water. Taking advantage of a veneer of sand that blots up moisture, Hopi farmers plant their crops in floodplain plots and alluvial fans, on spring-fed terraces at the base of the mesas and in dunes that store rainfall on top of the mesas. Dancing the rain down out of the sky, and using water-conserving cultural practices, the Hopis continue to harvest this thirstland. In a desert basin to the south, Hohokam farmers gouged out 30-foot-wide diversion ditches with digging sticks as early as 300 BC. Over time they built more than 1,750 miles of irrigation canals, but around AD 1400 the Hohokam abandoned their villages, and the land reverted to desert. Possible reasons for the exodus include salinization, competition for water, and withdrawal outstripping streamflow – problems that still plague irrigated agriculture.

In 1847 a new age of irrigation began when Mormon pioneers diverted a mountain brook to flood a patch of barren Salt Lake Valley. Throughout the West, Mormon farmers set patterns of cooperative irrigation aimed at optimal use of scarce water. In the seventeen contiguous western states, precipitation averages less than half that in the eastern states. Yet the West accounts for 80 percent of the nation's water consumption, with farmers by far the biggest users, thanks largely to the Bureau of Reclamation, which has converted deserts into high-yielding gardens by building 700 dams, 16,000 miles of aqueducts, and 35,000 miles of irrigation ditches. Irrigation has supplanted dry farming in much of the central and southern Plains, where water from the Ogallala Aquifer greens corn and alfalfa for fattening beef cattle.

The far West draws most of its water from surface sources. With the best dam sites used, U.S. reservoir capacity stands at about 500 million acre-feet. Underground aquifers supply most of the water for the Great Plains, where storage in the Ogallala under Nebraska alone is an estimated 2,100 million acre-feet. The water is free for the pumping, and 150,000 wells puncture the Ogallala. Trapped for thousands of years with little replenishment, this "fossil water" risks depletion, and even where recharge occurs, water is often pumped out faster than aquifers refill, causing water tables to fall. In 1986 hydrologists estimated that the Ogallala's southern extension was already half depleted. And in many western river basins, withdrawals total 60 to 80 percent of streamflow, acerbating salinity. California's diversion of northern waters to Central Valley farms and southern urban areas has dried up one lake and shrunk another.

Competing for water, farmers find themselves pitted against claimants for industry, energy, wildlife, recreation. Urban developers are paying twenty, even fifty times more than rural users can afford for water. And proving that "water runs uphill to money," some farmers are selling their water rights. Nonagricultural water demands are expected to triple in the next fifty years. But agriculture uses such a big share of our water, that small savings in irrigation can free water for large increases in nonfarm uses. Conservation measures can yield more water per dollar invested than can construction projects.

One of the wonders of water is that it is never irretrievably lost. Though we use it in myriad ways, we can never use it up. But neither can we make any more of it; in one form or another, its quantity is fixed. Nor can we ever rival such natural phenomena for storing and transporting it as snowpack, lakes, and aquifers; rivers and springs; clouds and rain. It is vain to seek ceaselessly to expand supplies of water by some technological sleight of hand. Rather we must moderate demand. We must use our water resources wisely, or else learn their value the hard way. When the well is dry, Benjamin Franklin admonished, we know the worth of water.

Left: Two swathers, cutting and windrowing alfalfa near Moses Lake, Washington, mow their way around a water wheel of modern agriculture, a center-pivot irrigation system half a mile in diameter. What makes the harvest possible in this semiarid region are electric pumps delivering copious amounts of well water through rotating sprinklers, producing four cuttings of hay per year.

Overleaf: Cumulonimbus cloud releases a localized torrent of rain near Third Mesa, Arizona, where mean annual precipitation is only ten inches, less than half of what it takes to grow a good corn crop in Iowa. But Hopi Indians have grown corn here for hundreds of years using techniques of rainfed and runoff farming.

● *Right*

SPRINKLER ARMS a quarter mile long
spray water at the rate of 1,000 –
1,200 gallons per minute, each
watering about 130 acres of alfalfa
in a single revolution, near Moses
Lake, Washington. Amid shifting
dunes of sand augmented by
considerable ash from the 1980
eruption of Mount St. Helens, the
irrigated fields hold the light and
sandy windblown soils in place.
Center-pivot systems cost from
$45,000 to $60,000, but they help
make economic use of marginal
land. Farms around Moses Lake
raise sixty-six different crops,
including potatoes, sugar beets,
peas, and various kinds of hay.
About 85 percent of the water used
in the Western states is used to
irrigate crops.

● *Right*

GREEN CHIPS in a high-yield game,
circular fields of irrigated corn
cover the landscape in north
central Nebraska. With a few
exceptions, each center-pivot
irrigation system fits neatly in a
quarter section, four in a mile-
square section bounded by roads.
Some farmers operate only two or
three pivots but others manage as
many as 120. Using new low-
pressure systems, they apply about
an inch of water per revolution, ten
to twenty revolutions per season,
depending on need. Average
annual precipitation for the area is
about twenty-two inches, near the
minimum for corn production, and
even small variations can mean a
bumper crop or crop failure.
Moreover, high-yield farming
requires not only an adequate but
a timely supply of water. Drawing
on the Ogallala Aquifer, the
Nebraska farmer has water on tap,
though he must pump it from
depths of 120–400 feet or more.

● *Left*

OASES IN the desert, center-pivot systems suck up well water to rain on alfalfa, wheat, vegetables, and cotton, near Yuma, Arizona. Some sophisticated systems can square a circle or skip an impediment. But sprinkler irrigation is declining in the area for a number of reasons: The high evaporation rate and low-holding capacity of the soils require irrigation every five to seven days during the growing season – 3,000,000 gallons of water on an acre of hay. Constant use exacts its toll on equipment designed for supplemental irrigation. The local groundwater has a high salt content. And local gophers find the sandy soil, well-stocked with roots, an ideal habitat.

● *Left*

THE COLUMBIA River supplies water for center-pivot irrigation in Washington's Horse Heaven Hills, with no appreciable effect on the river's flow. About 150 systems in the area produce mainly wheat, corn, and barley; one grows wine grapes. The sprinkler arms travel over tall crops as well as short ones, and, constructed of hinged sections, they negotiate uneven terrain with ease.

● *Overleaf*

FLUSHING SALTS from contoured fields in California's Coachella Valley, farmers flood fallow land with from one to five acre-feet of water, sometimes more than they use to irrigate crops. Soluble salts leach through the soil into buried drains. Caused mainly by use of saline Colorado River water for irrigation, salt buildup can significantly reduce soil productivity.

● *Above*

● *Right*

SALT-RAVAGED CORNFIELD in Grand Valley, Colorado, suffers from groundwater salinity caused by infiltration of irrigation water. The problem is due to poor drainage and the high salt content of the water from the Colorado River.

TRACTORS PULLING tandem scrapers level a field for uniform irrigation in California's Imperial Valley. They will come within 0.05 foot of the specified grade thanks to laser equipment, which made the checkerboard taking elevations.

● *Above*

MOBILE "BIG guns" shoot jets of water 150 feet, each firing 600–750 gallons of water a minute on rows of ripening tomatoes, near Homestead in southern Florida. Involving no costly installation, this method of irrigation is attractive in an area where most vegetable growers lease land under short-term contract. The water is drawn from wells, six wells to ten acres; the water table is high and easily accessible. Irrigation frequency depends on the evapotranspiration rate of the crop, which in turn depends on daily temperature, wind velocity,

relative humidity, and precipitation. During the growing season, tomatoes might have to be watered every five to seven days for 30–45 minutes. Wind can distort the spray pattern, changing it from a circle to an ellipse – the stronger the wind, the longer the ellipse – with the result that some areas of a field may receive no irrigation. Because the water is flung so far through the air, much is lost to evaporation before it reaches the ground. In addition, because the pattern is so large, water is wasted when it is sprayed off target, on neighboring fields, roads, or even houses.

Although average annual rainfall is relatively high in the South, seasonal distribution is often not optimal for crop production, and farmers have turned increasingly to irrigation, which can boost yields dramatically. In Florida, all horticultural crops are irrigated – tree fruits and nuts, vegetables, and ornamentals. The long growing season and water-responsive crop mix help justify installation costs, and some growers are now investing in micro-irrigation systems, notably drip irrigation, which is more water-efficient and can deliver fertilizer periodically in small amounts.

● *Above*

SALT DEPOSITS streak irrigated cropland in the Coachella Valley in California. The dark areas are saturated with water; when the water evaporates, dissolved solids are left behind. Brought from the Colorado River via canal, the water used for irrigation in the valley has a high salt content, about 1.1 tons of salt per acre-foot of water. Soil salinity can be aggravated by high evaporation rates in the hot desert climate; slow water intake due to compaction; breakdown of chemical fertilizers; too little irrigation, which fails to flush salts from the root zone; and too much irrigation coupled with poor drainage, which can raise the water table and bring salts to the surface. Land leveling, artificial drainage, and leaching help prevent high salt concentrations, which can inhibit plant growth. To ensure good yields, choice of crop is also important. Salt-sensitive crops include sweet corn, alfalfa, carrots, and grapes. The most salt-tolerant include dates and figs among the tree fruits; barley, sugar beets, and cotton among the field crops; and tomatoes, asparagus, and broccoli among the sixty vegetables grown in the valley.

● *Overleaf*

CRISP PATCHWORK of crops – dark green alfalfa, light green wheat, fallow fields ready for cotton – covers an arid plain in central Arizona first irrigated by Hohokam Indians more than 2,000 years ago. The present Salt River Project – drawing on streamflow and groundwater – was the first project built under the 1902 Reclamation Act. Over decades of greening the desert, water levels near Phoenix have dropped as much as 400 feet, and competition for water has risen. But use of groundwater is now declining as the Central Arizona Project comes on line, bringing Colorado River water hundreds of miles over the mountains.

● *Right*

ANCIENT OXBOWS and meanders
emerge from alfalfa fields near
Yerington, Nevada. Over millennia
the channels were filled with sand
and gravel, materials that do not
retain water as well as the
surrounding soils. In the fall of a
second year of drought, the alfalfa
plants have died in many of these
areas. During the winter, cattle and
sheep graze these fields. Come
summer, high-protein alfalfa hay
grown for California milk cows is
irrigated with water from the
Walker River; a canal cuts across a
field at lower left. Parallel ridges in
the fields control the flooding that
occurs in border irrigation.
Angling through the fields, a
subsurface drainage ditch prevents
a waterlogging and salt buildup.

 ● *Right*

ALAMO RIVER snakes across the
ordered agricultural landscape of
California's Imperial Valley. The
riverbed was greatly enlarged
when floods sent the entire flow of
the Colorado River into the valley
from 1905 to 1907, creating the
Salton Sea in this desert basin
some 200 feet below sea level,
where the average annual rainfall
is 2.91 inches. Long since tamed,
the waters of the Colorado now
travel via the 82-mile-long All
American Canal and a 1,675-mile
network of subsidiary canals to
irrigate nearly 500,000 acres of
cropland; and the Alamo serves as
a central drain in a 1,457-mile
system that collects surface runoff
and subsurface drainage and
discharges it into the Salton Sea. In
1988, land that would not be worth
a nickle without water produced
$976,481,000 worth of agricultural
products. The top commodity was
lettuce, with a gross value of
$199,206,000, followed by cattle,
alfalfa, cantaloupes, carrots, and
twenty-nine more multimillion-
dollar crops.

● *Left*

SKIRTING THE escarpment of Yuma Mesa, a canal conveys water from the Colorado River to irrigate the Yuma Valley around Somerton, Arizona. The earthen channel delivers the water to lateral canals, which route it in turn to irrigation ditches, where turnout structures divert the water to flood fields that are pool-table flat, with no fall in any direction. Some seventy-five crops are grown in the valley, including cotton, wheat, alfalfa, lettuce, cauliflower, and watermelons. Bermudagrass and many other crops are grown for seed, a specialty favored by the area's low rainfall and twelve-month growing season. Citrus crops – grapefruit, tangerines, Valencia oranges – thrive on the sandy soils of the mesa.

● *Left:*

CANALS AND highways cross and converge in the Rio Grande Valley, near Las Cruces, in southern New Mexico. Some eighty miles upstream, the river is dammed and its water stored in Elephant Butte Reservoir for release to this area for irrigation during the summer. Average annual precipitation in the state ranges from six inches in arid basins to thirty-five inches in high mountain areas. Most runoff to refill reservoirs comes from spring snowmelt, plus summer thunderstorms. In years when snowfall in the north is deficient, water for distribution to the central and southern parts of the state is inadequate, and agricultural users supplement it with groundwater. As intricate as the irrigation network is the web of local regulations, state laws, national legislation, eight interstate compacts, and three international treaties that govern the use of New Mexico's limited water resources.

● *Above*

SLICE OF high-yielding cropland is
sacrificed to a farmstead near
Yuma, Arizona. But the farmer has
made up for the lost acreage in the
adjacent field by adding a fold-out
extension to the sprinkler arm to
reach into the corners.

● *Right*

SUCKED DRY, the Rio Grande in
central New Mexico won't flow
again until runoff season. Virtually
all freshwater on the surface of the
state has been appropriated.
Agriculture accounts for more than
90 percent of water withdrawals.

188

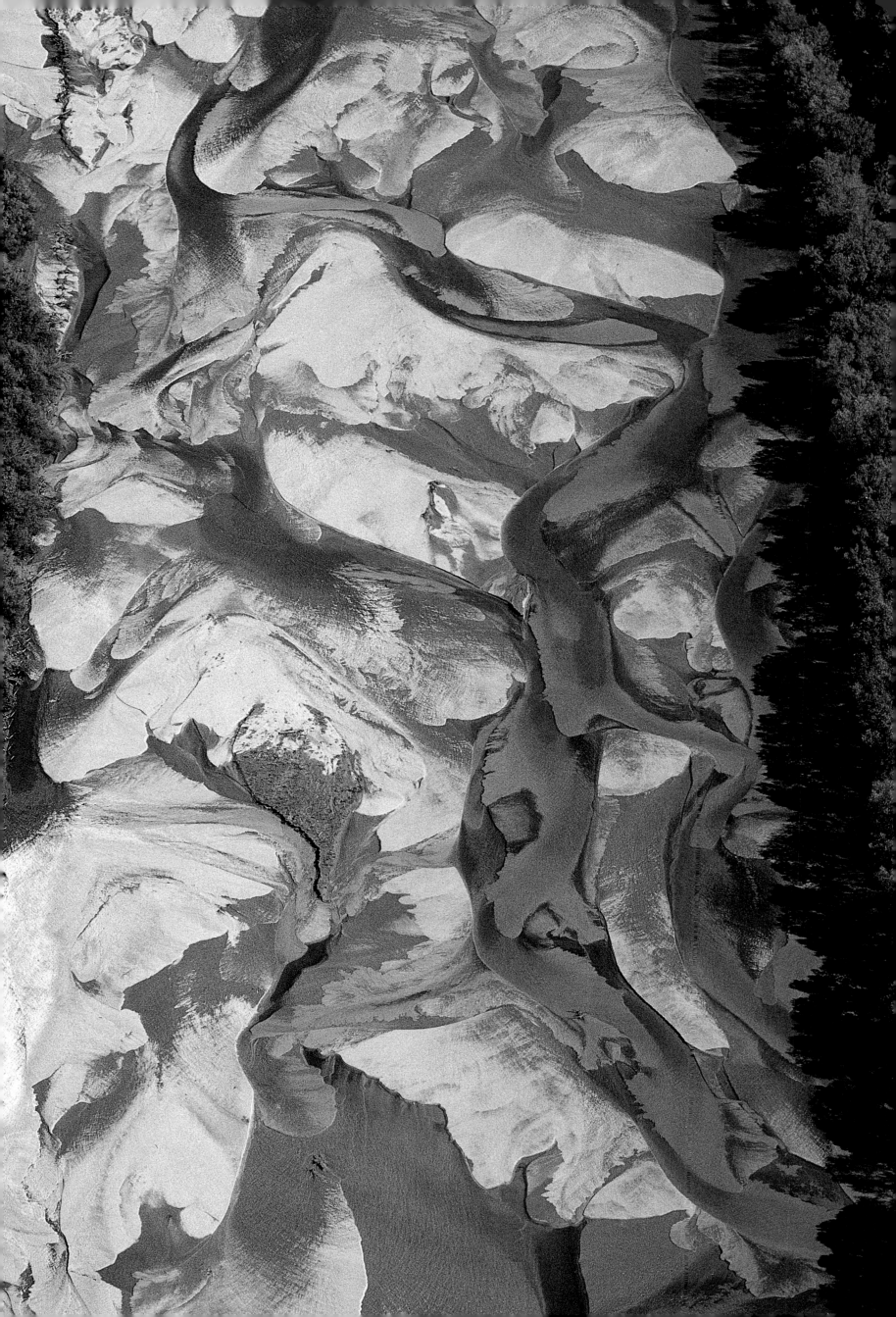

LOSING GROUND

Created by generations of farmers clearing forests, plowing prairies, draining swamps, and watering deserts, America's cropland represents an incalculable investment of labor and capital. Yet even before we had won independence we were losing farmland to urban pressures. "We live too thick," complained a husbandman in mercantile New England in the 1700s. With industrialization, factory villages encroached on fields, and water impounded to run the mills flooded upstream bottomland. Abandoning hard-won ground in the East, pioneers pushed westward, and as long as the American frontier expanded, cropland gains outran losses. Since about 1910, however, land planted to crops, or temporarily idled, has remained fairly stable at around 400 million acres. But much of the land harvested today is not the same as that harvested yesterday. Every year, roughly a million acres of cropland are transformed into roads and reservoirs, residential developments, commercial centers, industrial parks, and other uses. Offsetting these losses, a million acres are brought under cultivation elsewhere, often in more fragile areas.

As cities spread, sprawl, and stretch out along highways, developers outbid farmers and ripple effects reach ever deeper into the countryside. And the more crowded our cities become, the greater the demand for a recreational hinterland; our most urbanized regions have become our most forested – with farmland a two-way loser. When city folk move in next to farmers, the newcomers complain about manure odors, cows that moo in the night, and hay wagons that slow commuter traffic. Farmers complain about pets harassing livestock, teenagers riding dirt bikes across new-sown fields, and skyrocketing taxes. Some have become strangers in their own land as nonfarmers, who now make up 85 percent of the rural population, "zone out" agricultural activities. Digging in, farmers and their supporters have countered with zoning laws that restrict development, state or county purchase of development rights, preferential tax assessment, and "right-to-farm" laws exempting farmers from nuisance suits.

Much of the land lost to urban growth is prime cropland, and some unique ecological niches are in jeopardy. Artichokes thrive only in one small coastal area of California, now at risk from urban incursion, and smog from the Los Angeles basin is spilling into the desert valley where virtually all our date palms grow. Competition for limited water supplies also threatens to curtail cultivation. Seeking to eliminate critical groundwater overdrafts and allow for urban and industrial growth, Arizona has mandated that irrigated cropland – 1.3 million high-yielding acres of former desert – be reduced by nearly 40 percent by the year 2025. Millions of acres of wetlands have been drained to become some of our best cropland – marshes in the Florida Everglades, "prairie potholes" in the upper Midwest, bottomland hardwood forests in the Mississippi Delta. Today, wetlands are protected against conversion to crop use by the "swampbuster" provision of the 1985 farm bill, as are highly erodible drylands by the "sodbuster" provision.

As land available for cropland expansion dwindles, saving farmland in the city's shadow becomes a matter, not just of preserving a piece of our past, but of conserving a valuable productive resource. More than one-quarter of American farms are located in or on the fringe of metropolitan areas. On less than a sixth of the nation's farm acreage, they produce almost a third of U.S. farm output in terms of value. Unable to halt the urban tide, many farmers have turned to small-scale, intensive agriculture, growing high-value crops and livestock: melons, berries, and tree fruits; vegetables, herbs, and greenhouse products; Long Island ducklings and Belgian rabbits. Commercially oriented, they often sell directly to consumers at farmers' markets and roadside stands, and to hotels, restaurants, and discriminating supermarkets. And farmers and their families are taking advantage of metro-area opportunities for education, recreation, and off-farm employment. In a world where the traditional boundaries between urban and rural have blurred, such innovative patterns hold promise that farmers and farmland may become a lively, lasting, and integral part of the metropolitan scene.

Left: Circling their wagons, mobile-homeowners put their mark on erstwhile cropland near Phoenix, Arizona, with an eighteen-hole golf course, parks, lakes, and other amenities of a retirement playground. Set in the desert Southwest, Phoenix has swelled to become the ninth biggest U.S. city, intensifying competition with agriculture for land and especially water.

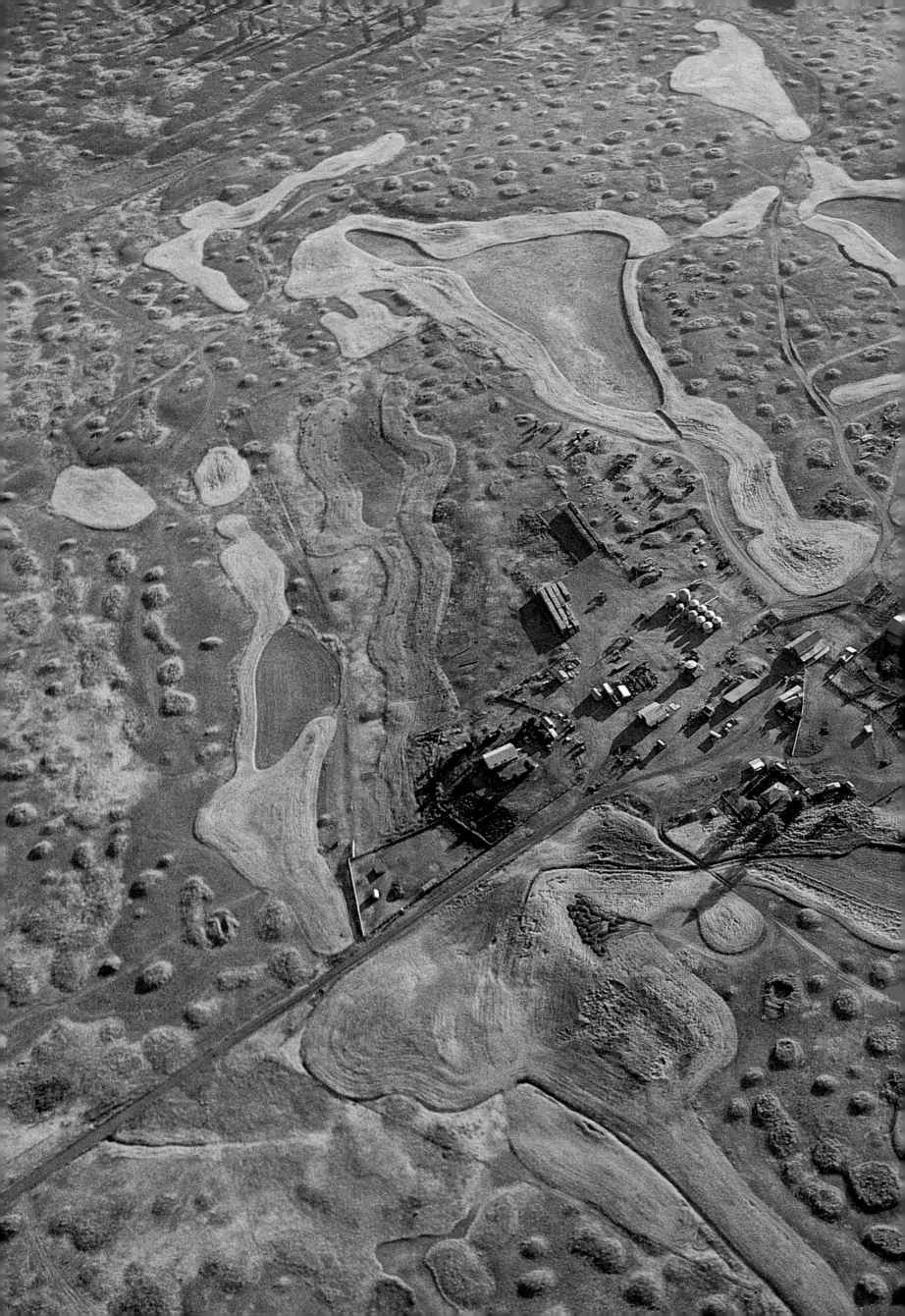

● *Previous pages*

"SCABLANDS" IN southeastern
Washington provide grazing for
cattle on the hummocks, with
grain production on larger patches
of soil; intervening shallow soils
support little vegetation. Some say
the mounds are the work of
gophers. Most geologists believe a
huge flood sculpted the scablands
when a glacial ice dam broke and a
vast lake suddenly emptied,
sweeping away a thick mantle of
loess and scouring the underlying
basalt. As prime farmland comes
under urban pressure, agriculture
tends to expand into areas that are
less fertile, more arid, more prone
to erosion, or ecologically valuable
in their natural state.

● *Above*

GOUGED OUT by the same colossal
deluge, potholes and rock basins
near Othello, Washington, have
changed little since the Pleistocene,
neither eroding nor filling in with
sediment in the very dry climate.
The scarred basalt provides very
poor drainage, and the soil
covering is scanty. An attempt at
center-pivot irrigation with waste
water from a potato-processing
plant appears to be a failure,
yielding only a growth of weeds or,
at best, pasture grass. The adjacent
feedlot is a more successful
venture, fattening 10,000–12,000
head of cattle every four months.

● *Right*

KEEPING THE Badlands at bay, good
conservation practices on the part
of farmer William Sturtevant
prevent "the breaks" from eating
into his fields near Wauneta,
Nebraska. Even more important,
his conscientious use of terracing,
contour stripcropping, and crop
rotation combats the invisible but
far greater threat of wind erosion,
holding soil losses to less than one
ton per acre per year, whereas
losses could exceed twenty-five
tons on unprotected cropland.
Such measures can be costly, but
without them, cultivating fragile
areas can be even more costly
in terms of erosion.

● Right

WATERFOWL REFUGE near Enderlin, North Dakota, bears the imprint not of Big Bird but of a bulldozer rooting out stands of overgrown cattails to improve a marsh as a breeding area. The region produces fully half of North America's duck crop. As agriculture has lost ground to development, farmers have brought new lands into production, often destroying such wildlife habitat. More than 50 percent of the Dakotas' pothole wetlands have been drained, as have 99 percent of Iowa's marshes. Now wetlands are protected, but the competition is not one-sided, and croplands remain vulnerable, not just to urban encroachment but to conversion to parks, preserves, forests and resorts.

● Right

FLOURISHES OF an interchange ornament former farmland, where State Highway 41 passes over U.S. Highway 44 and veers northeast into Oshkosh, Wisconsin. The highway construction boom of 1956–1975, which involved acquisition of 1.8 million acres as right-of-way, continues to echo in the development of adjacent land. Enterprises around this 25-year-old interchange include, clockwise from lower left: a farm implement dealership; the home of a retired farmer, in the triangle; a packaging plant under construction; a recreational vehicle center, south of the cemetery; and an outdoor drive-in being demolished to make way for "Aviation Plaza," a mall that takes its name from the land to the south, owned by the Winnebago County Airport and the Experimental Aircraft Association, which annually hosts one of the nation's largest air shows. The EAA Museum is just out of the photo to the right. No trace remains of working agriculture.

● *Left*

INTERSTATE 29 cuts through the dark rich soils of onetime prairie, near Elk Point in southeastern South Dakota. Besides physically appropriating productive acreage, highway construction can disrupt tillage and drainage patterns, and can make access difficult when it splits a farmer's property. Although local impacts can be enormous, on the national scale cropland losses to development are not large. Competition is often keenest, however, for level, well-drained soils, which are prized not only for growing strawberries but also for building shopping malls. And conversion to built-up uses tends to be irreversible.

● *Left*

PLANTED AMID fields of hay and corn and stretches of forest, a residential subdivision adds to the increasing acreage devoted to nonagricultural uses in the farm country of southeastern Pennsylvania. Lancaster County is blessed with fertile soils, a favorable climate, skilled farmers, and proximity to major markets, but as population has grown over recent decades, more than an acre of farmland has been taken out of production for each new resident; in 1987 alone, an area nearly twice the size of Lancaster City, the county seat, was approved for housing, commercial, and industrial development. Pennsylvania has now passed farmland preservation legislation. In addition, concerned Lancaster citizens, seeking ways to retain productive farmland yet accommodate growth, are urging local officials to permit higher density development and to promote in-filling, or use of undeveloped land in areas already urbanized.

● *Above*

"Nothin' to stop the wind but a bob-wire fence," they say in Texas. Still, a Panhandle farmer has roughened his field with a sandfighter in a bid to prevent erosion, caused when high winds meet up with smooth, sandy soils.

● *Right*

Surrounded by tangled wetlands, a field in southern New Jersey has been drained, diked, and planted to cabbage or iceberg lettuce. Of the original 215 million acres of wetlands in the contiguous United States, fewer than half remain.

OUR AGRARIAN ROOTS

Only when we gave ourselves outright "To the land vaguely realizing westward,/ But still unstoried, artless, unenhanced," did we become Americans, poet and one-time farmer Robert Frost reminded us at the inauguration of John F. Kennedy, evoking the primal bond between land and people. Without the land to give scope to dreams and put its continental imprint on questing immigrants, there would be no Americans. Nor would there be an America without people who had the vision and vigor to create a nation from a vast, amorphous wilderness. The New World challenged, offered possibility and promise. Again and again, a new breed took up the challenge. The American farmer, remarked Ralph Waldo Emerson, "changes the face of the landscape. Put him on a new planet and he would know where to begin."

Inexorably, the land shaped us as a people and we the people shaped the land, in patterns of interdependence that live on in the landscape and persist in our culture. In the ordering of fields and pastures, farmsteads and rural communities, we sense some larger order, expressing attitudes and aspirations that spring from the seedbed of shared experience. Truths we hold to be self-evident – the right to "Life, Liberty, and the pursuit of Happiness" – are rooted in our agrarian past. The author of those words, Thomas Jefferson, held freehold farmers to be the best of citizens: "They are the most vigorous, the most independent, the most virtuous, and they are tied to their country, and wedded to its liberty and interests by the most lasting bonds." The loved look of farmland reflects the covenant between farmers and the land they till.

The first colonists fondly remembered well-tended landscapes shaped over centuries by Old World peasants, but the reality of a harsh, raw land where they had to start from scratch forced radical adaptations. To survive in heavily forested New England, settlers planted Indian crops in hastily cleared plots. Gradually stumps were removed, stones stacked to form walls, and a patchwork of small, scraggly fields emerged. But farmers could not take advantage of advances in agricultural equipment on hilly, boulder-strewn terrain, and when the Erie Canal opened up new areas in 1825, many Yankees moved west to lands "ripe for the plow." Southern planters improvised a shifting agriculture devoted to tobacco and cotton, crops that were highly profitable but soon exhausted even the most fertile soils. As old fields were abandoned, new fields were brought into production, largely by African slaves using hand tools. An agricultural reformer deplored the "butchered" landscape of Virginia around 1800: "Farm after farm had been worn out, and washed and gullied, so that scarcely an acre could be found . . . fit for cultivation."

In the middle colonies, Germans, Swedes and Finns, who had experience in settling virgin forest, introduced efficient methods of land clearing and log construction. While preserving many Old World traditions, they made innovations in agriculture, improving plows, breeding strong draft animals, rotating crops of grain with grass and clover to maintain soil fertility. As they pushed inland, Pennsylvania farmers carved out large fields, which met their own needs in abundance and left surpluses for market. These patterns not only laid the foundation for permanent, prosperous agriculture in the region but served well for farmers spreading west through the Ohio Basin. Beyond the Mississippi, however, pioneers would have to develop new techniques for breaking sod and settling the treeless prairie. Swelling the ranks of American-born frontier farmers were immigrants bringing new contributions.

Underlying agricultural patterns are survey patterns. In the eastern English colonies, the "metes and bounds" system used natural boundaries or lines connecting natural landmarks, such as trees and boulders. Land-grant recipients commissioned surveys to include fertile soils, springs or other coveted features within their allotted acreage, producing a crazy quilt of fields – and a legacy of boundary disputes. In areas of the Mississippi Valley subject to French influence, the "long lot" system divided land along rivers into narrow strips, giving water frontage to a maximum number of landholders. In 1785 Congress created national order out of regional chaos by dividing the federal domain into huge square "townships" aligned foursquare with the compass; six miles on a side, each township was subdivided into thirty-six square-mile sections. With minor corrections for the curvature of the earth, this rectangular grid was extended from horizon to horizon as new lands were annexed. A 160-acre quarter section became the basic land claim under the Homestead Act of 1862; to promote settlement of the arid West, the Desert Land Act of 1877 offered homesteaders a full 640-acre section. The rigid system riled farmers when it narrowly deprived

Left: Neighbors, closer than most in rural Wisconsin, have just planted the prairie loam to corn for livestock feed. The dark brown of the soil reflects its organic content, the legacy of grasslands that once covered some 140 million acres. Red corn cribs bulge with ears from prior harvests. Across the road, silos tower over the barnyard, and flowering crab, ash, maple, and boxelder trees bring spring to the dooryard. Trees that grace old farmsteads often commemorate a birth, a marriage, a new undertaking, and some farm families still plant a tree to mark an auspicious event.

them of a stream or potential woodlot, but on the whole, the great checkerboard came to be accepted as giving full and fair measure, as evidenced in the popular expressions "square meal" and "square deal."

The fundamental unit of American agriculture, however, is the family farm. Pioneer farming was of necessity a family undertaking, with husband, wife and children from an early age contributing to the common effort to put food on the table, clothes on their backs, and shelter over their heads. The work ethic, on which Americans pride themselves, harks back to the dedication of farm families turning virgin land into a productive enterprise. Self-sufficient rural householders spun wool, tanned leather, cut timber for building, ground grain for bread, butchered their own hogs, preserved their own fruits and vegetables. No pioneer family could do without apples. The pick of the crop was eaten fresh; bushel after bushel was cut and dried and hung from the rafters, or boiled down into apple butter; windfall apples were pressed to make cider – sweet in the first flush of autumn, taking on zing as it hardened; extended fermentation produced cider vinegar for farmwives' pickles, while distillation yielded applejack. For want of nearby craftsmen, the American farmer became a jack-of-all-trades. "Most of us are skilful ... in mending and making whatever is wanted on the farm," wrote J. Hector St. John de Crèvecoeur around 1780.

Following Old World prototypes, New England colonists had settled in villages with their fields clustered round about, but in the middle colonies and regions to the west, where settlement was scattered, an enduring American community pattern emerged – the rural neighborhood. For each family, the farmstead was the hub of a neighborhood that encompassed the farms within a day's walk or ride and various points of convergence: the local church, the school at the crossroads, the mill on the river, perhaps a hamlet with a few stores and craftsmen. Neighborhood comprehends not just a relationship to place but a bond among people that comes close to kinship; it implies an inherent equality, congeniality and willingness to lend a helping hand. William Cooper Howells recalled the neighborliness of Ohio pioneers in the early 1800s: "Their houses and barns were built of logs, and were raised by the collection of many neighbors together on one day ... The men understood handling timber, and accidents seldom happened, unless the logs were icy or wet or the whisky had gone round too often." Roof raising would be followed by supper and a dance.

But the central fact of our agrarian past was not frontier settlement but westward expansion and continual resettlement. Pennsylvania was the test plot for pioneering in Ohio, and Ohio in turn for territories beyond. From the outset, along the length of the frontier, farmers driven by a restless energy had moved west along parallels, repeatedly lured over the next horizon. Each new frontier – the coastal plain, the Piedmont, then beyond the Appalachians – offered new fields of opportunity. With the Louisiana Purchase from France in 1803, the young republic doubled its size; on concluding the $15,000,000 transaction, French foreign minister Talleyrand remarked to the U.S. envoy: "You have made a noble bargain for yourselves and I suppose you will make the most of it."

Among the waves of migrants that would flood across the Mississippi, the Missouri, the Great Plains and the Rockies were trappers and traders, explorers and surveyors, cattlemen and miners, but the quintessential frontiersmen – those who left their mark on the land and fulfilled Talleyrand's prediction – were the pioneers who transformed the wilderness into productive farmland. Whole families endured hardships they could not have imagined when they dared to undertake the adventure. Some pulled up stakes and moved on half a dozen times before settling on land from which they hoped to coax or coerce prosperity. The first home on the plains was often a dugout scooped out of a rise, walled off with sod and roofed over with prairie grass. Sustained by the optimism that gave them the courage to seek their destiny in America's uncharted vastness, polyglot pioneers emerged from the crucible of the frontier resourceful American farmers, as rugged as the land, their self-confidence buoyed by confidence that God was on their side.

Traits and values forged on the frontier – individualism, ingenuity, self-reliance; belief in hard work and fair play; love of family and community, home and the land; faith that dreams can become reality – strike chords in all of us. Although relatively few of us can trace our ancestry to the first settlers, or even to the nine out of ten Americans who lived and worked on farms at the time of independence, these national forebears live in our collective memory. And almost anyone whose American roots go back fifty to a hundred years is likely to have at least one homesteader in the family tree. For hundreds of thousands of immigrants, from earliest times until most recently, the bond with America through working the land has been real and immediate. Today only 2 percent of the population live on farms, but none of us are far removed from the land. Our agrarian past is not dead agricultural history but a living cultural heritage, a pattern of ideas and ideals that mutually define America and Americans.

Right: "Oldest Family Farm in America," near Dover, New Hampshire, is operated by the eleventh generation of Tuttles to farm the land since John Tuttle claimed it with a grant signed by King Charles I in 1632. Far from tradition bound, however, the family pioneered some of the newest U.S. farming patterns, and today Tuttle's Red Barn sells the entire output of the farm and greenhouses — cut flowers, nursery products, strawberries, herbs, and vegetables — to a growing urban market.

● *Previous pages*

● *Above*

WINDBREAKS OF close-set evergreens
shelter a prairie farmstead south of
Cannon Falls, Minnesota, where
Norwegian immigrants settled in
1878. The property has been
farmed by the family ever since,
today by Franklin Hessedal, a
great-grandson. The barn, now
over a hundred years old, is still
sound, its frame all pegged and
doweled – not a nail in it. The
curving grassed waterways are
marks of good stewardship of the
soil, as are the different crops
planted in alternate strips: Bright
green sweet corn is rotated with
soybeans, mostly just emerging, a
legume that will produce nitrogen
to nourish next year's corn crop.

CHRISTMAS TREES await harvest in
Waushara County, Wisconsin,
where the sandy loam soils are
ideal for growing such Yuletide
favorites as Scotch pine, Douglas
fir, and blue spruce. Six to eight
feet tall, nine-year-old Scotch
pines, the industry staple, stand in
serried ranks separated by fire
lanes. Consumers seek trees that
are straight and symmetrical, and
do not shed their needles. Growers
look for fast, uniform growth and
resistance to pests. Despite such
advances as planters that put in
1,000 seedlings an hour, growing
Christmas trees remains labor-
intensive. Often trees must be
trimmed to achieve a conical

shape, and before harvest they are
sprayed with green stain to
enhance their color. Come
November, mature trees are cut
with a chain saw, then baled and
bound and shipped to market. One
grower claims Christmas trees are
depression proof – during the
worst of the Thirties, families still
bought a tree and one present per
person. The threat today comes
from artificial trees, that have
captured more than a third of the
market. But there's more than
meets the eye, and it's not likely
that forever-green plastic will ever
fool the nose with that pungent
piney smell that is as seasonable as
sleighbells.

● *Above*

Avenues of loblolly pines converge on the red brick manor house, built in 1838, of Jamaica Point Farm, near Trappe, Maryland. The road on the right leads to Jamaica Point, a sandy spit in the Choptank River, which flows into Chesapeake Bay. The point probably got its name from the region's brisk West Indian trade in colonial days. Trade was also brisk with England, the principal export being tobacco. Along the 2000-mile shoreline of Chesapeake Bay and its navigable tributaries, large and small planters sought waterfront locations where they could load hogsheads of tobacco directly onto sailing vessels, and Jamaica Point Farm

was doubtless tilled during tobacco's heyday in the 1700s. Tobacco growers were mobile farmers, forever abandoning worn-out fields only to clear new ones. In the tidewater regions of the Chesapeake, however, large plantations became anchored to a "big house" and prosperous planters became attached to a luxurious lifestyle. When the land could no longer produce tobacco, sedentary planters turned to other crops, notably wheat and other grains, but also crops of slaves to be sold to growers who had migrated to labor-short inland areas. The builder of the manor house, William B. Hughlett, farmed

part of his land but profited mainly from its timber, building a lumber business and constructing sea-going sailing ships at Jamaica Point. The present owner grows corn and soybeans on 350 tillable acres, just planted in this late spring picture, but he has made most of Jamaica Point Farm and its long riverfront a waterfowl sanctuary, maintaining ponds and leaving acres of standing corn for migrating and overwintering geese and ducks.

● *Right*

Dairy farm in New York's Hudson River valley cycles forage and grain through cows to make milk to fill the shelves of metropolitan supermarkets. In an exercise pen, Holstein cows cluster round the feed trough; a mature animal will consume about fifty pounds of feed, dry weight, and thirty gallons of water daily. To satisfy their appetite, corn harvested green from the fields fills three silos, and hay is stored in gambrel-roofed barns at the end of the long dairy barn. In addition, grain concentrates help boost output. Jutting out from the dairy barn is the milking parlor, which the cows visit twice a day. A tanker truck collects the milk daily and delivers it to a processing facility.

● *Right*

Flying farmer Vern Ramesbotham grows corn and soybeans near Elk Point, South Dakota. He uses his 1946 Aeronca Chief, a "high-wing tail dragger," to go to Sunday morning flight breakfasts. He has removed the wings of another vintage aircraft for recovering. A self-styled "talented mechanic" but a procrastinator, Vern buys used equipment at auction and much of it that litters his farmstead is not junk, he maintains, but "unfinished projects." Bargain farm machines include a 1975 blue Ford tractor, parked near his house, and a forty-year-old green John Deere combine, parked next to a wooden grain bin. He bought the red Chevy truck, in the driveway island, in mint condition in 1973 for the equivalent of 300 bushels of soybeans. Vern uses the old cars scattered about for storage, since he has no farm buildings. If he built any, he claims, the county would just raise his taxes.

● *Left*

RED BARNS and tall silos once formed the infrastructure of dairy operations, but today Snake Hill Farm, near Middleburg, Virginia, raises beef cattle. The farm produces large amounts of hay, but good forage is not the only reason for the rise of the cattle business in the South. It requires less labor than almost any form of agriculture. Whereas dairy cows must be watered, fed, and milked every day of the year, beef cattle can fend largely for themselves. And wealthy investors have been attracted to the businesss not only by tax advantages but by the "gentleman squire" aura that surrounds the breeder of prime cattle and purebred horses.

● *Left*

THREE DOUBLE-width poultry houses near Dover, Delaware, produce about 75,000 tender young broilers, raised from day-old chicks, every seven weeks – more than half a million birds a year. Agriculture is Delaware's biggest industry, bigger than the state's three chemical giants – DuPont, ICI, and Hercules – combined. And poultry accounts for more than 80 percent of agricultural sales. The principal crops are corn and soybeans, which go mainly into carefully formulated chicken feed along with other grains plus vitamins and antibiotics. The DelMarVa Peninsula, comprising Delaware and parts of Maryland and Virginia, is well located to take advantage of the growing demand for poultry in a huge metropolitan market. With Americans' heightened health consciousness, per-person consumption of chicken overtook that of beef in 1989.

● *Above*

FALL DETERMINES the colors at
Schartner Farm Market, in Exeter,
Rhode Island – bright orange
pumpkins; creamy, gold, coral, and
crimson hardy fall mums; snowy
white and deep pink flowering
cabbage. Here Schartner Farms sell
to local residents the fruit,
vegetables, flowers, and other
nursery stock that they grow in
their fields and greenhouses.
Direct marketing, through retail
outlets or pick-your-own farms, is
increasingly popular with both
growers caught in the urban
squeeze and consumers with a
range of tastes, for exotic or
organic produce, or simply for
good, fresh food and flowers.

● *Facing page*

HARVEST BOARDS groaning with the
bounty of the Pickaway Plains form
part of an eight-block midway
during the Circleville Pumpkin
Show, held each October in
Circleville, Ohio. To the right of the
"pumpkin tree" are the Largest
Squash prize-winners in various
categories. Besides pumpkins,
gourds, and squashes, dozens of
other vegetables are displayed and
judged, as are crafts, baked goods,
pickles, flowers, pets, and babies.
There are hog-calling and pie-
eating contests, parades, clowns,
and carnival rides. A high-schooler
reigns as Miss Pumpkin Show, and
a first-grader as Little Miss
Pumpkin Show.

● *Overleaf*

WINTRY PANORAMA of the American
Midwest was shot from an airliner,
probably over Kansas or Nebraska
where I-70 or I-80 slices across the
landscape. Drifted snow and the
low angle of the winter sun
enhance the topography, revealing
complex patterns of nature,
culture, and agriculture. The
turbulent forms of nature are
evident in watercourses,
drainageways, and rugged terrain.
Embedded in the rectangular
matrix of sections and quarter
sections is a cultural mosaic of
independent, landowning farmers.
Joining nature and artifice,
agriculture's handiwork appears in
great irrigated circles and
contoured terraces.

210

INNOVATION AND ANCIENT WISDOM

Older than civilization, agriculture has evolved over thousands of years as humans have sought ways to induce nature to do their bidding. Native Americans walked softly, their farming marked by a filial reverence for nature. Hopis see the sun as their father and the earth as their mother, whose milk is the corn that nourishes them; successful agriculture depends on honoring this affinity. European settlers saw nature as a wilderness to conquer. Cutting down forests and plowing up grasslands, advancing farmers replaced whole populations of plants and animals with species of their own choosing; they changed the contours of the land and altered the course of rivers. They achieved unprecedented abundance, often at the cost of the environment. An old woman of the Wintu tribe mourned their abuse of nature: "How can the spirit of the earth like the White man? . . . Everywhere the White man has touched it, it is sore."

Innovation overtaking ancient wisdom is a recurring theme in agriculture; in America, a series of technological revolutions has accelerated the process. In 1800 a farmer wrestling with a wooden plow and sowing seed by hand put in 344 hours of labor to produce 100 bushels of corn, harvesting twenty-five bushels per acre. By the late 1930s mechanization had cut a farmer's work to 108 hours for 100 bushels of corn, but yield per acre remained about the same. By the late 1980s a farmer with high-tech equipment worked only three hours to produce 100 bushels of corn, and yield per acre had soared to 118 bushels, thanks mainly to improved seed, irrigation and chemical pesticides and fertilizers. Along with these changes, the diversified family farm has increasingly been converted into a factory-style operation devoted to one or two crops, sometimes planted to thousands of acres, and the farmer has become a mini-industrialist wrestling with credit and cash flow, guided by values that presume bigger is better. And agriculture is only one component of the mega-industry known as agribusiness. On the input side are industries that provide the supplies and equipment farmers need to produce crops and livestock. On the output side are industries that process, package and distribute farm products. This sector claims most of the consumer's food dollar, leaving less than thirty cents for the farmer.

Today biotechnology promises to make plants and animals healthier and hardier, more nutritious and more productive. Agricultural scientists have long since replaced natural selection with selective breeding; now molecular biology has enabled them to supersede evolution through genetic manipulation. This revolution could restructure American agriculture. In the 1990s almost a million farms are expected to succumb to economic pressures and consolidation, reducing farm numbers to 1.2 million by the year 2000. At the top, 50,000 superfarms — profiting from economy of scale, equipped with the latest technology, often managed by hired experts — will produce 75 percent of the nation's food and fibre. At the base, a multitude of small farms, many of them part-time enterprises, will specialize in fresh produce. Unable to compete successfully, moderate-size owner-operated farms are projected to decline to 6 percent of the total.

Many farmers are now heeding a call to counter-evolution. Caught in a spiral of costs for fertilizers and pesticides, they ruefully recognize that they are fertilizing weeds as well as their crops and creating havens for insects in vast monocultures. Off-farm, agricultural chemicals have caused water pollution and raised concerns about food safety. The industrialization of farming has not only degraded soils but eroded the rural community. And alarms have been raised about tampering with genetic codes. Triumphs of technology are not always victories for agriculture or culture. Today state universities and the U.S. Department of Agriculture are pursuing research on alternative farming methods, and a National Academy of Sciences study has concluded that chemical-free farming can be just as productive and more profitable than farming requiring expensive chemical additives. By employing systems that work with, not against, nature — conservation tillage, crop rotation, biological pest control — farmers could restore economic and ecological health to large segments of agriculture. By whatever name — alternative, low-input, or organic — farming that relies less on purchased inputs and more on natural processes requires the dedication of a farmer walking his acres. The "best of all fertilizers," wrote a down-home sage in the *Prairie Farmer* in 1851, is "the owner's foot."

Left: Shocks of corn follow contours laid out by an Amish farmer in north central Ohio, as do cut stalks waiting to be bundled, and standing corn not yet harvested.

Overleaf: Spread like a hooked rug over the Palouse hills, near Pullman, Washington, a 200-acre experimental farm is patterned with test plots of wheat, peas, lentils, forages, and hay crops.

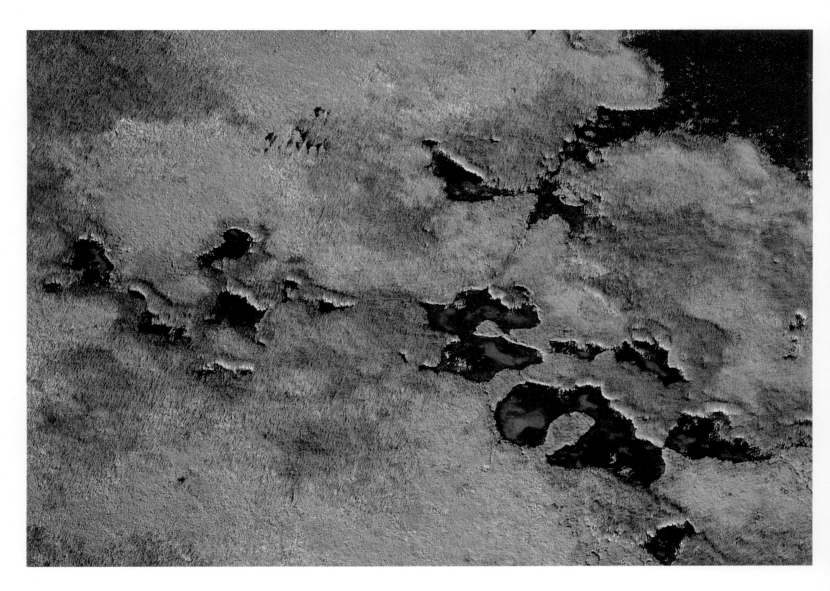

● *Above*

MOTTLED LIME-GREEN distinguishes wild rice, the only native North American grain, from darker alder, willows, and other bog plants that thrive on the shores and in the waters of Leech Lake, Minnesota. In late July, when this picture was taken, this aquatic grass, which grows in shallow water, begins to flower, and by the end of August the first grain is ripe. The traditional Indian method of harvesting, which can continue into October, is done by two people in a canoe: a poler who navigates through the rice, and a knocker who uses one cedar stick to bend the rice stems over and another to knock the grain into the boat. Two experienced ricers can harvest about 200 pounds of rice in a day. The method is sufficiently inefficient to ensure future crops, since much of the grain falls into the lake. The Ojibway Indians also seeded new marshes, but they never domesticated wild rice. Not until about 1970 was a variety developed with nonshattering seeds, which remain on the stem for machine harvesting. Today, combines harvest as many as 10,000 pounds of wild rice an hour from high-yielding California paddies. And lake-grown, hand-harvested grain now accounts for less than 10 percent of the wild rice crop.

● *Right*

GARDENING IN the sand, Hopi farmers grow corn, beans, watermelons, and peach trees by making efficient use of scarce water, near Third Mesa, Arizona, probably the longest continuously cultivated area in the United States. Drought-resistant Hopi corn sprouts from seeds planted as deep as a foot or more, where the soil remains damp; plantings are widely spaced and staggered in alternate years to give roots access to adequate moisture. Large stones, brush fences, or standing stalks are used to protect seedlings from drying winds.

● *Above*

● *Right*

FROM ASPARAGUS, in the uppermost plot, to tennisball lettuce, next to the pavilion, Thomas Jefferson's kitchen garden at Monticello, Virginia, now restored, served as much as a laboratory as a source of food for the table.

"GARDEN OF EDEN" was created by mystic Adam Purple on a demolition site in New York City. He fertilized his mandala of roses, vegetables, and herbs with droppings from Central Park's horses.

● *Right*

Deep in a bin brimming with corn, a front loader feeds the golden harvest to an auger, which lifts it into a storage elevator owned by a farmers' cooperative at Elk Point, South Dakota. Four out of five commercial farmers are co-op members, notably dairy farmers, grain and soybean producers, and nut, fruit, and vegetable growers, who join together to market more than a quarter of the raw farm products. Some of the nation's top agribusiness enterprises are cooperatives, and some of the most familiar names on grocers' shelves – Land O'Lakes, Sunkist, Ocean Spray, Blue Diamond – are a sign not only of quality but of the cooperative spirit of farmers.

● *Right*

Rows of used combines in a salvage yard in Leesburg, Indiana, can be a barometer of the farm economy. In good times, sales of new equipment go up, in lean times, the salvage business is brisk. These self-propelled combines were bought all over the Midwest. The parts are sold at 50 percent of new price – by self-service. Farmers go into the yard and take the part they need off the equipment themselves. Fueled by the agricultural export boom, sales of new farm machinery peaked in 1979, at $11.7 billion, only to fall until less than a quarter as many tractors and combines were sold in 1986. During those years of farm crisis, demand for used equipment drew down dealers' inventories, then new sales picked up again in the late 1980s as net cash farm income rose to record levels.

222

● *Left*

BULGING STORAGE bins north of Tower City, North Dakota, constitute a prairie farmer's bid for prosperity. Crops stored away when prices were low will be cashed in when prices rise. With proper management, the grain in these bins – probably wheat, barley, and/ or soybeans – could be stored for several years. The bins range from about 5,000 to 15,000 bushels in capacity. The long Quonsets would hold much more, or they may house machinery. Much of the world's reserve of grain is stored on American farms.

● *Left*

NEW WINE in new bottles – the output of grapes pressed around the state is stored in huge tanks at the Gallo Winery at Modesto, California. Like giant thermos bottles, the tanks are lined with stainless steel or glass, and coated with four inches of polyurethane foam. The red tank has been scraped and is being refurbished; after new foam is applied, the tank will be repainted. Stored at low temperatures and blanketed with nitrogen to seal it from the air, the wine can be held for years. The rectangle at top left is the roof of a vast cellar where some premium wines are aged in oak. All Gallo's wines – screw-cap table wines and cork-finished varietals – are blended and bottled at Modesto. The administration building, at upper right, is the operational hub of the largest wine-maker in the United States and perhaps the world. The Gallo brothers, Ernest and Julio, took up wine-making as Prohibition ended, learning the trade from pamphlets put out by the University of California.

● *Above*

● *Right*

UNION STOCK Yards of Omaha, Nebraska, founded in 1883, received 4,971,760 head of livestock in 1967. Today packing plants are built near feedlots, and much of this property is now a shopping center.

BRUSH CORRAL protects sheep from nighttime raids by coyotes and bobcats on the Zuni Indian Reservation in New Mexico; by day they are herded from area to area. Zunis reserve almost a third of their sheep for festivals.

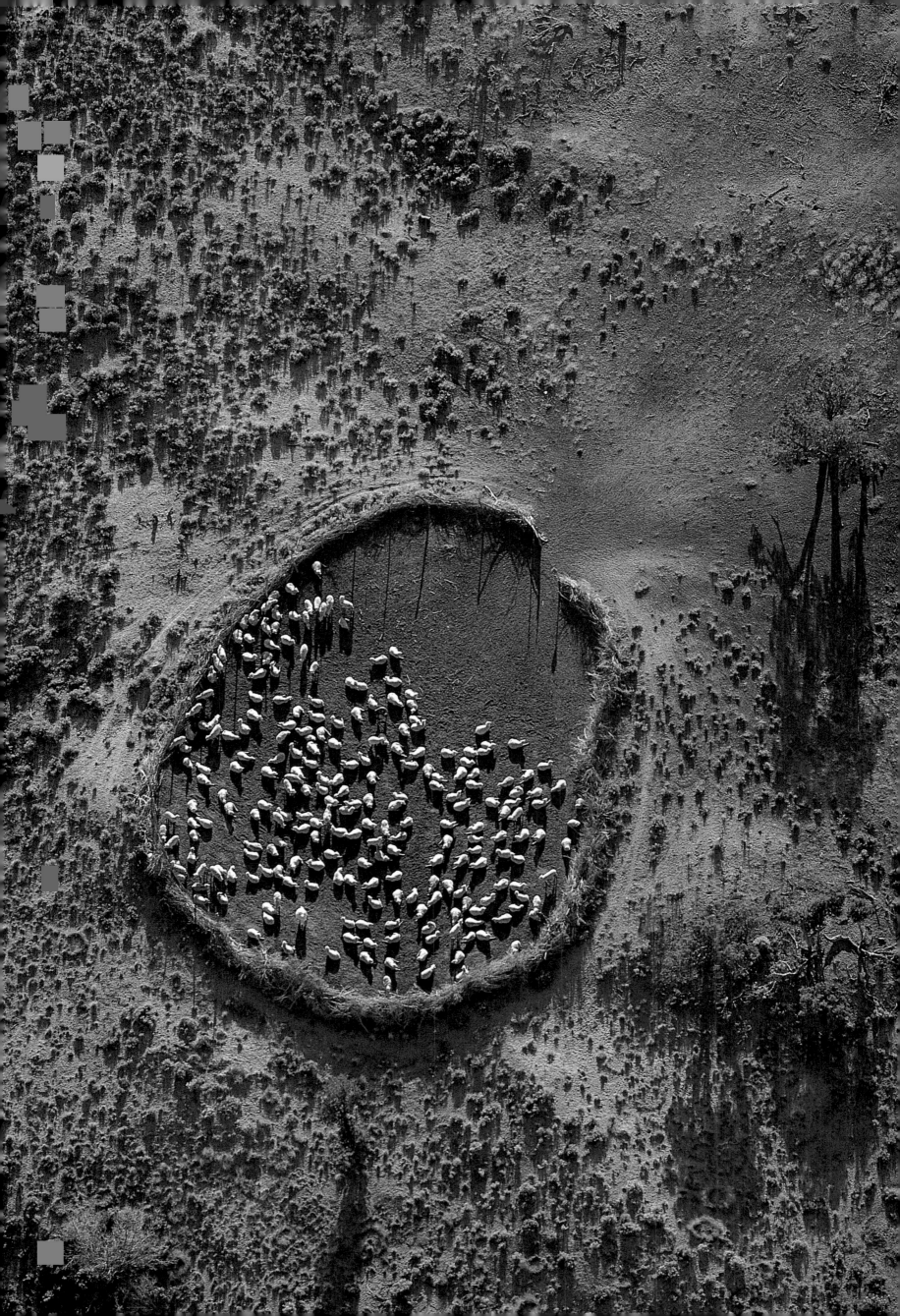

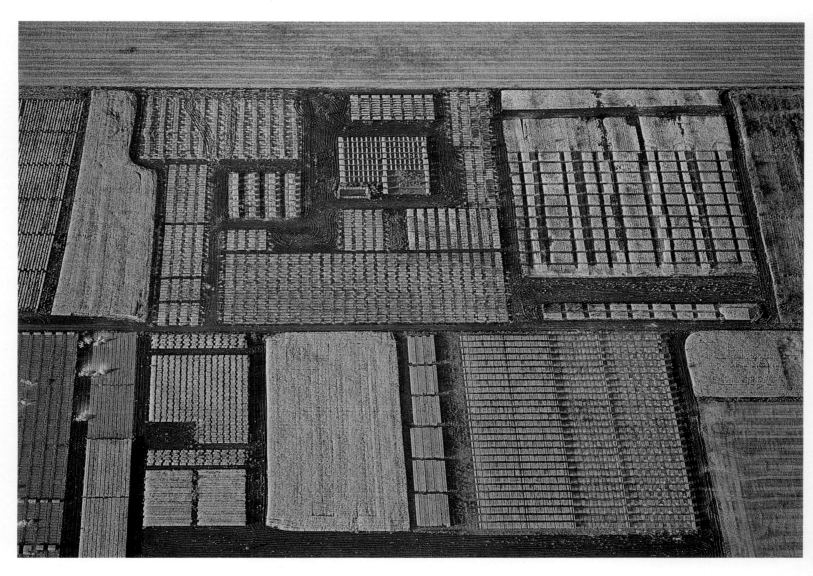

● *Above*

BREEDING GROUNDS for hundreds of
kinds of grain, the North Dakota
State University–Prosper
Agricultural Research Plots are
meticulously laid in the Red River
Valley, near Amenia, North Dakota.
Each of the small blocks contains a
different strain or variety of wheat
or other grain; the large blocks are
seed increase fields. Across the top
band, left to right, are a field of
durum wheat, breeding nurseries
of winter wheat and golden barley,
and a field of green flax; and across
the bottom, a durum nursery and
fields, a half-harvested nursery of
spring wheat, and more durum,
mostly harvested.

● *Facing page*

GIANT TIC-TAC-TOE tests the effects of
shelter on crop production at the
10,000-acre University of Nebraska
Field Laboratory, near Mead,
Nebraska. The windbreaks are
made up of Austrian pine, Eastern
cottonwood, and Eastern redcedar;
the fields in the 40-acre grid will
be planted in winter wheat.
Another experiment will compare
the flammability of various native
grasses growing in plots divided
into small squares; prescribed
burning will test ignition and fire
spread at different seasons.
Reducing the incidence of wildfires
along railroad rights-of-way would
help preserve wildlife habitat.

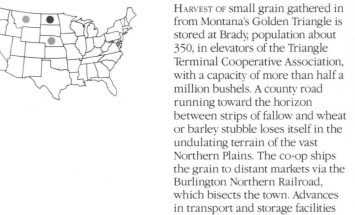

● *Overleaf*

HARVEST OF small grain gathered in
from Montana's Golden Triangle is
stored at Brady, population about
350, in elevators of the Triangle
Terminal Cooperative Association,
with a capacity of more than half a
million bushels. A county road
running toward the horizon
between strips of fallow and wheat
or barley stubble loses itself in the
undulating terrain of the vast
Northern Plains. The co-op ships
the grain to distant markets via the
Burlington Northern Railroad,
which bisects the town. Advances
in transport and storage facilities
opened the way for even the
remotest regions to compete
successfully in the grain trade.

FARMER AS ARTIST – ARTIST AS FARMER

A rt imitates nature, said the classical philosophers; modern critics contend that art expresses human creativity. In either case, the farmer is surely an artist and good farming is art, whether beauty inheres in the object or in the eye of the beholder. For what could be more creative than the farmer's collaboration with nature in the miracle of life and the glory of growth? The intrinsic beauty of farmland lies in fertility and abundance. Its visual delights spring from the colors and textures of soils and crops, the contours and shapes of conservation and irrigation, the patterns created by planting and harvesting – all changing with the seasons and weather, time of day and stages of growth. Since prehistoric times, nature's materials – soil, sunlight, stones and boulders, growing things, flowing water – have inspired artists the world over to fill earth's empty spaces with images. Often working in the same media as farmers, today's land artists also profit from modern farm technology, using tractor-drawn disks and nitrogen fertilizer to create giant field graphics.

The look of landscape turns on one's perspective, physical and mental. Those who work the soil cherish personal visions of their creations. For the beholder, a view from above is most likely to apprehend the complexity and scale, the tensions and rhythms of the land as seen by the inner eye of the farmer/artist. Expansive aerial vistas also coincide with our cultural perspective, for space, more than time – the sweep of geography, more than of history – dominates the American psyche. Celebrating the prairie landscape as "first-class *art*," "nourishing . . . to the soul" as well as the body, Walt Whitman groped for words to express its grandeur, "that vast Something, stretching out on its own unbounded scale, unconfined, which there is in these prairies, combining the real and ideal, and beautiful as dreams." A vertical vantage point extends the horizontal: length and breadth reach from horizon to horizon. And though height is compressed, a bird's-eye view enhances depth. The landscape unrolls like a giant palimpsest, revealing the layered poetry of geology, ecology and human activity, the overlapping, interlacing calligraphy of nature and culture, art and artifice, agriculture and applied geometry.

At the opposite extreme, Hamlin Garland affords an intimate, down-to-earth view of the prairie, seen by boys lying on their backs in the "green deeps" of an Iowa wheatfield on a midsummer day: "We trembled when the storm lay hard upon the wheat, we exulted as the lilac shadows of noonday drifted over it! We went out into it at noon when all was still – so still we could hear the pulse of the transforming sap as it crept from cool root to swaying plume . . . our hearts expanded with the beauty and the mystery of it – and back of all this was the knowledge that its abundance meant a new carriage, an addition to the house or a new suit of clothes."

Immersion in the landscape gives immediacy to farming's sounds and smells, adds detail, intensifies color. An elevated perspective opens our eyes to farming's dazzling patterns – the dark mahogany scrollwork of freshly turned earth, the velvety green tracery of tender shoots, the crisply etched graphics of sunstruck stubble, kaleidoscopic mosaics of cropland and woodland. The rhythm of the seasons is marked not only by changing hues but by the cycle of chores moving in synchrony with nature's cycle from germination to maturation, from dormancy to resurgence and renewal. Responding to the hand of the husbandman, the life-giving earth takes on a life of its own, as in Frank Norris's evocation of planting time in California in 1901: "The great brown earth turned a huge flank to [the sky], exhaling the moisture of the early dew . . . The rain had done its work; not a clod that was not swollen with fertility, not a fissure that did not exhale the sense of fecundity . . . the land was alive; roused at last from its sleep, palpitating with the desire of reproduction . . . offering itself to the caress of the plow, insistent, eager, imperious." The powers that the land elicits from the plowman are various – artistry, potency, skill, industry, love. But the farmer does not differentiate: all are part of the art and care that he brings to that most basic of human tasks, the growing of food and fiber. And in the results of his efforts, beauty and bounty are confounded in patterns of plenty.

Left: Deftly maneuvering a tractor, artist Stan Herd lowers and raises his plow to outline a rotund vase and draw laciniated leaves and petals in a clover field near Eudora, Kansas. The lighter brown background was plowed a week earlier. Later in the season, "Sunflower Still Life" – an earthy emulation of Van Gogh's favorite theme – will be enlivened with emerald green soybeans and bright yellow blooms of the Kansas state flower.

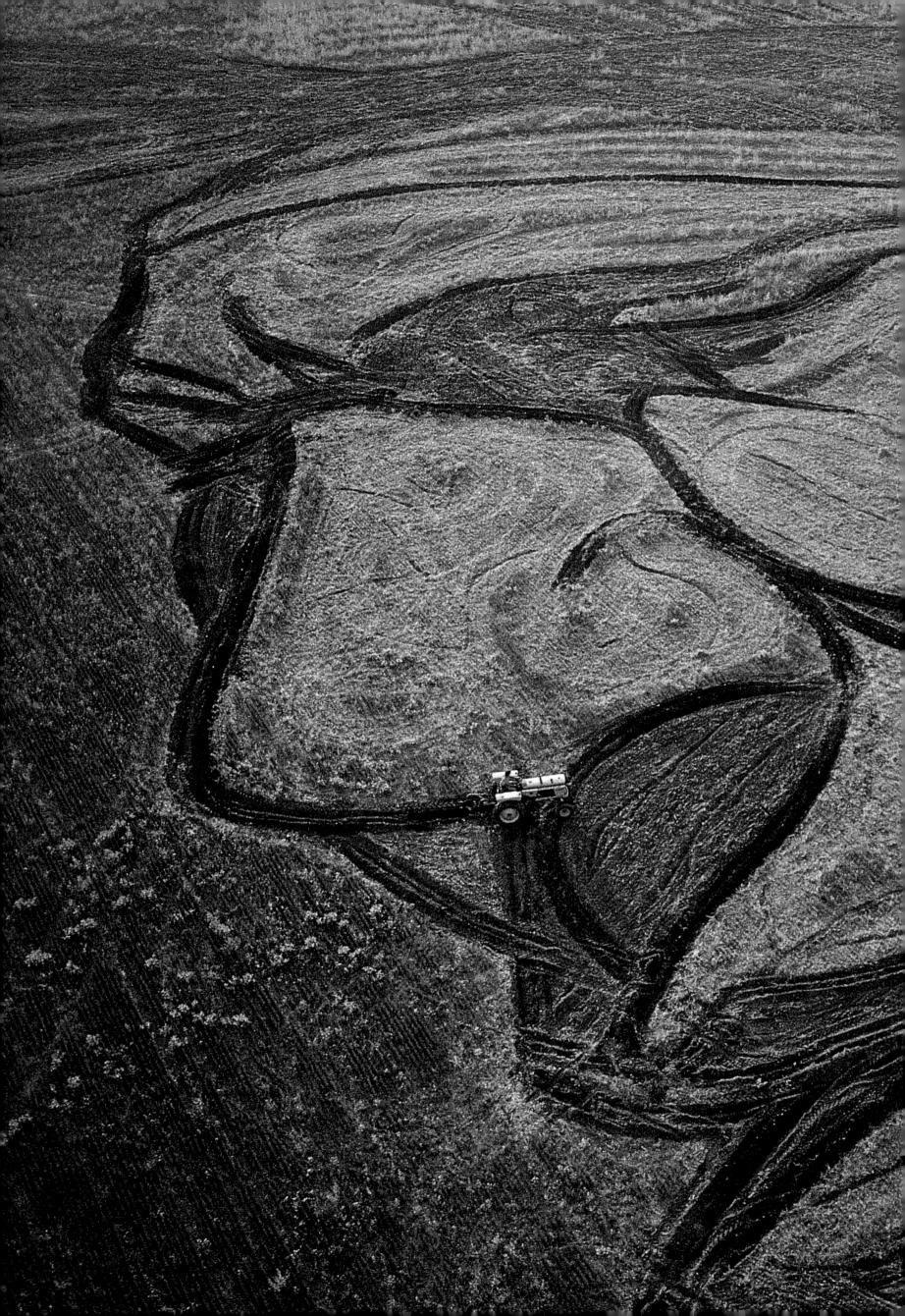

● *Previous pages*

TURNING UP the dark earth with a
two-bladed plow, Stan Herd incises
a proud Native American nose in a
field of wheat stubble near
Oskaloosa, Kansas. To render the
thirty-acre line drawing of a Sac
and Fox Indian, he used a variety
of implements like different
pencils: three different tractors,
two different mowers, and three
different plows, ranging from six to
twenty feet in width. The grandson
of homesteaders, Herd has
incorporated the work of his
farmer forebears into his own
work as an artist to achieve a new
form of creative expression
through working the soil.

● *Above*

"PORTRAIT OF Saginaw Grant,"
completed, measures about 1,000
feet from the top of Grant's
feathered headdress to his
neckerchief. To transfer realistic
detail to that scale, Herd first
makes a sketch on paper and
overlays it with a grid; he then lays
out a corresponding grid in the
field with numbered flags at each
intersection. When late summer
rains greened the portrait with
volunteer wheat and prairie grass,
Herd plowed the outlines and
shadows afresh and, to create
highlights, shaved areas of the face
clean with a mower.

● *Right*

CRUSHED PEPSI and Coke cans litter
the combat zone in "The Ottawa
Beanfield Cola Wars," staged by
Stan Herd in a soybean field near
Ottawa, Kansas. Against a
backdrop of plowed earth, lime
forms the logos on cans of soybean
stubble and – in a new departure
for crop artist Herd – people
provide the color, red or blue,
wearer's choice. Because showers
discouraged some fair-weather
warriors, Herd, atop a cherry
picker, had to compose the piece
one can at a time.

● *Right*

LAYERING PATTERN upon pattern, a
caterpillar tractor turns heavy
muck soil with a thirty-inch disk to
prepare a seedbed for vegetables,
south of Lake Okeechobee in
Florida. The horizontal pattern was
made when the field was laser
leveled. The vertical lines and
loops result from installation of a
system of subirrigation, or
reversible drainage: A tractor
pulled a torpedo-shaped "mole"
through the soil, leaving a six-inch
tunnel, or mole drain, running
from one canal almost to the other;
the tractor then raised the mole,
backed around in a loop, and
repeated the process. By raising or
lowering the water level in the
canals, the farmer can irrigate the
field during dry periods or drain it
after heavy rains.

● *Right*

BUNCHES OF straw punctuate a
harvest pattern near Chester, in
Montana's Golden Triangle.
Following the land's contours, a
combine operator narrows the
island of standing wheat with each
sweep; as the cut wheat is
threshed, a straw buncher attached
to the combine catches the straw
and unloads it periodically. The
bunches may be left for livestock to
forage, or more likely, the straw
will be baled and removed from
the field. The richest small-grain
growing region in the state, the
Golden Triangle encompasses
eight counties in the area between
Cut Bank, Havre, and Great Falls.
The gently rolling to nearly level
terrain, made up of glacial till
deposited during the last ice age, is
easy to farm in large blocks or
strips a mile long by forty rods
wide, for high yields of winter,
spring, and durum wheat, as well
as barley and oats.

● *Left*

INTRICATE WEB of barley near Paso Robles, California, unravels its secrets only to an expert eye. The alternating ribbons of ripening grain and bare ground are due, it seems, to a malfunction the farmer failed to notice when the crop was planted: one of the two seed drills towed by his tractor apparently jammed. The interlacing pattern comes from the farmer planting the far corners first, then planting the rest of the field across the slopes – both excellent erosion control practices. Beginning just to the right of the web's center, the farmer first made long loops to the left, to the bottom, and to the right, and then made the rounded triangles, working from the center outward.

● *Left*

DELICATE CALLIGRAPHY on a rain-drenched field in Kent County, Delaware, stems from some grubby agricultural activity, such as spraying pesticide or applying fertilizer. Reflected in the puddles, glints of the late afternoon sun and a clearing sky augur well for the corn crop, due to be planted within ten days. It takes 5,000 gallons of water to grow a bushel of corn, but for bumper crops precipitation must be timely. In the spring, rain is needed to recharge dry soils and provide moisture for germination, but May downpours can put a damper on planting. If fields are too wet to work, farmers fall behind schedule, and crops planted late risk damage from frost before they are ready for harvest. During the summer, corn draws on subsoil moisture with roots that go down four feet, but rain is required at the tasseling stage when pollination occurs. In the fall, heavy rains may again cause losses by delaying the harvest.

● *Above*

LIKE PAINT applied with bold
brushstrokes, vivid wild mustard
covers traces of cultivation near
Portsmouth, Virginia. The former
cropland may be fallowed in a set-
aside program or enrolled in the
Conservation Reserve Program.

● *Right*

FOCAL POINT of a black-and-white
engraving, a sinkhole cradles a
cluster of trees near Rochester,
Minnesota. Crosshatching the
winter fields are vertical lines of
fall tillage and, underneath, the
diagonal rows of last season's crop.

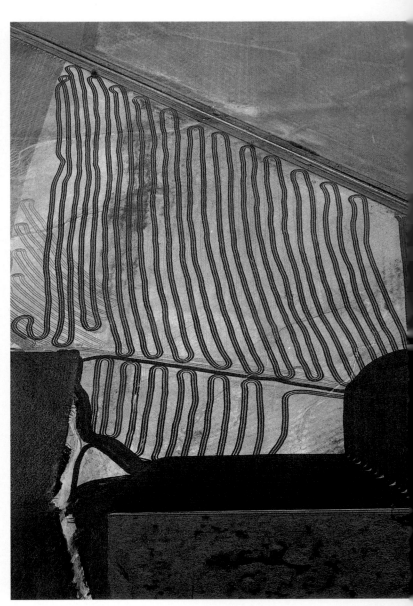

● *Top left*

ABSTRACT PATTERNS in western Nebraska – rectangles versus stripes represent opposite approaches to agriculture: block farming versus stripcropping. The rectangular design was created by a farmer tilling inward from the field's perimeter; to make a turn, he raised his implement, probably a plow, leaving untilled diagonals. On one side, he also raised his plow to leave a skipper or grassed waterway. The dark rectangle has yet to be plowed. Mottling occurs where wind or water has blown or washed away topsoil, exposing lighter colored subsoil; these erosion-damaged areas will be much less productive. The striped pattern consists of strips that are cropped every other year: The green strips are winter wheat, planted last fall and now growing in the spring; the golden strips are wheat stubble.

● *Top center*

FINE PEN lines from the air are rough ridges at ground level, turned up by a farmer in an effort to halt wind erosion, near Stafford, Kansas. The faint tracks at the left were made to combat an earlier wind; now the wind has shifted and so have the lines of defense. The broad brushstrokes at the bottom of the field are bands of dark soil turned up by a disk to prepare a seedbed.

● *Top right*

LIKE PAINT applied with a palette knife, crops and stubble and freshly tilled earth create a richly textured farmscape, west of McPherson, Kansas. The thinly spread green is winter wheat in spring growth stage. The thick green streak is a grassed waterway. The pale areas between the terraces appear to be wheat stubble. In the dark brown soil prepared for planting, the overlapping circles were made by a wide tillage implement pivoting to negotiate sharp turns.

● *Bottom left*

SCRAWLING THE landscape with the abandon of a child fingerpainting, a farmer seeks to cover the most ground in the least time as he wages a desperate battle to save his soil, near Hutchinson, Kansas. Unprotected by live vegetation or crop residues, the smooth whitish soil, dry and crusted, is highly vulnerable to wind erosion. Spurred to action by a rising spring wind, the farmer has used a disk or a field cultivator to turn up dark moist clods and roughen the soil to increase its resistance to erosion. The field will probably soon be planted to sorghum grown for grain or forage. The green field is well protected with a crop of winter wheat.

● *Bottom center*

PRECISE, PREMEDITATED pattern is the work of a farmer, near Mitchell, South Dakota, also intent on preserving his soil. The long, looping strips are probably sorghum planted on small-grain ground in late summer to help prevent wind erosion during fall and winter; in the spring the sorghum will be plowed or disked under to make way for a crop of spring wheat or other small grain.

● *Bottom right*

WHIMSICAL IMAGE – reminiscent of a painting by Klee, or a giant spider – appears to be a field in transition from rice to soybean production, near Lafayette, Louisiana. The roughly parallel lines trending horizontally are former rice levees the farmer has plowed out. He has also made a U-shaped pass in the old quarter drains to help dry out wet areas. After disking a broad band around the square perimeter prior to planting, he has switched to working the field in a circle from the center out.

● *Above*

STALKING THROUGH the wheatfields, a fallow field assumes a feline form along the ridges of Palouse country in southeastern Washington. Next year the pattern will be reversed – the cat will be tawny with wheat and the surrounding fields fallowed to store up moisture. A line of locust trees on the neighboring ridge breaks the force of winds that spill over the hills. The trees were planted in the 1930s by the Civilian Conservation Corps. Born of twin calamities, the Dust Bowl and the Great Depression, the CCC put jobless young men to work and carried out numerous erosion control projects and other conservation efforts.

● *Facing page*

STARTLED RABBIT rears out of the deep, dark loess soil of a fallow field near Lewiston, Idaho. The rabbit's coat of ripe winter wheat awaits the harvesting combine, which has already sheared the adjacent fields. The clump of trees that forms the rabbit's eye masks a rocky outcrop. The outline may follow the land's contours, with only the higher ground planted, but the pattern may also be determined by the farmer's need to limit planted acreage to comply with farm programs in order to qualify for government payments.

● *Overleaf*

"THY BELLY is like a heap of wheat set about with lilies" – Solomon's song of his beloved might also have celebrated the sensual curves of the grain-giving earth in California, where wheat and barley are grown on the crowns of rolling hills and in long narrow valleys at the northern approach to the Carrizo Plain. In May the pale gold winter wheat has just been harvested; these fields will be fallowed for eighteen months before being resown.

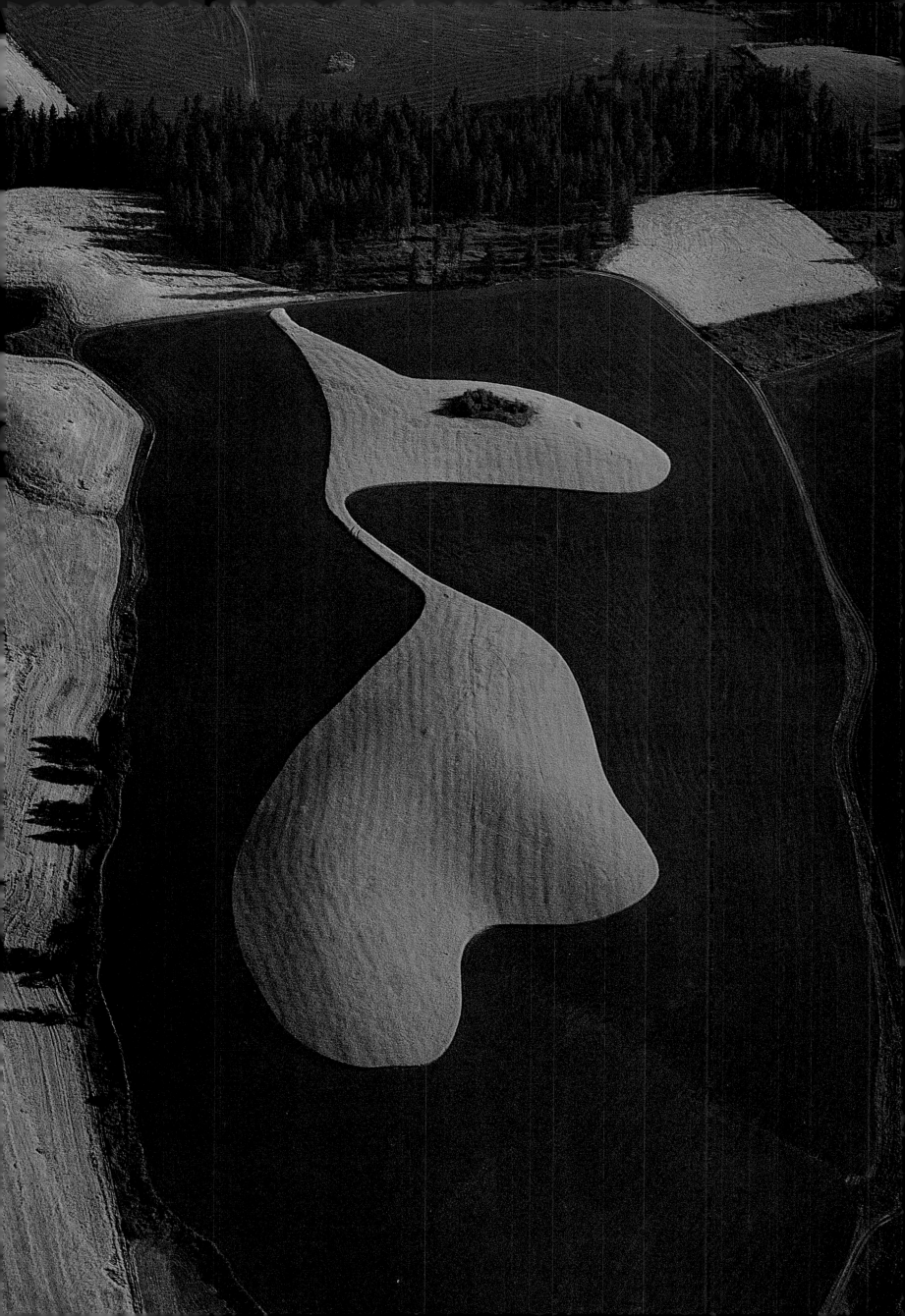

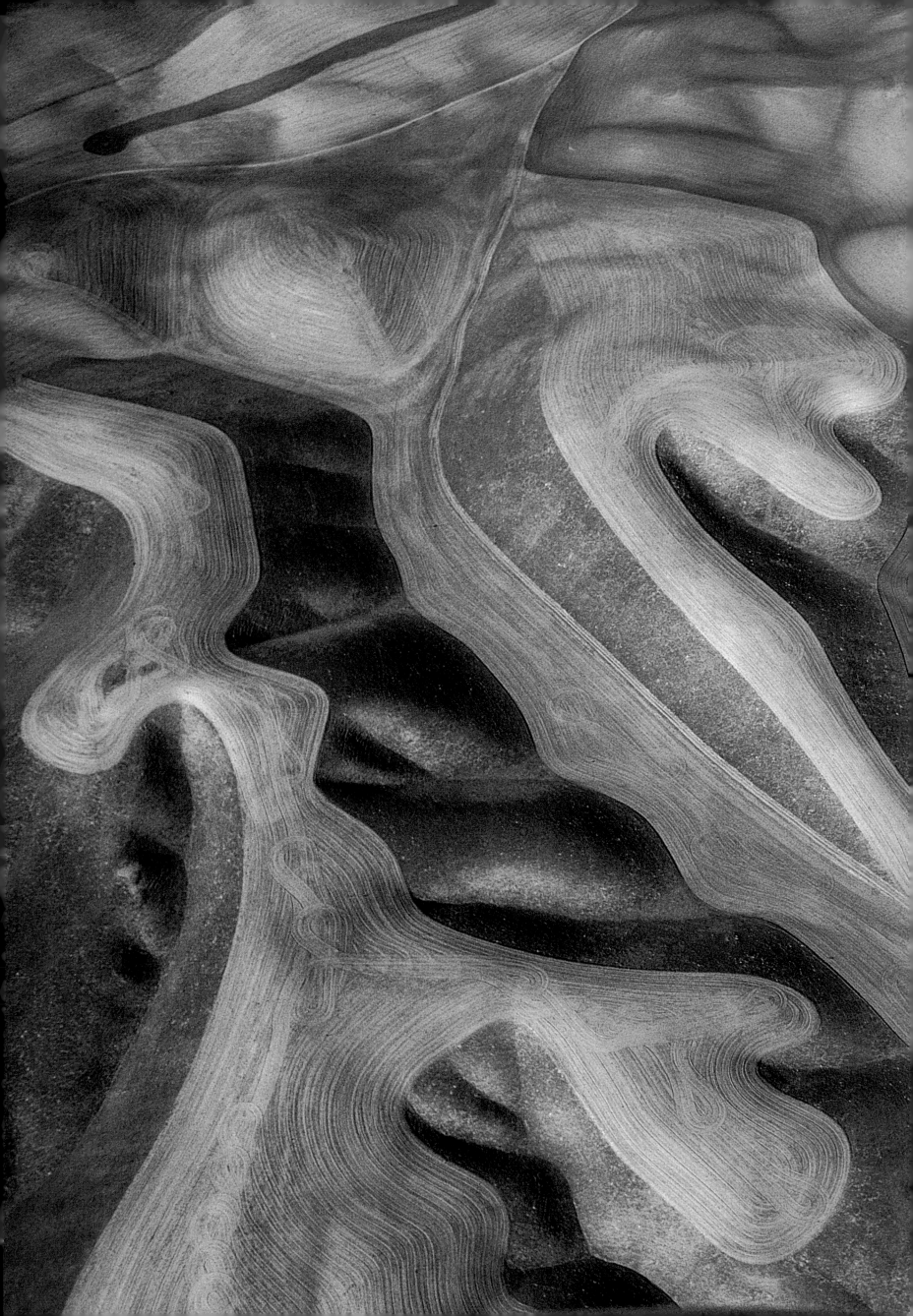

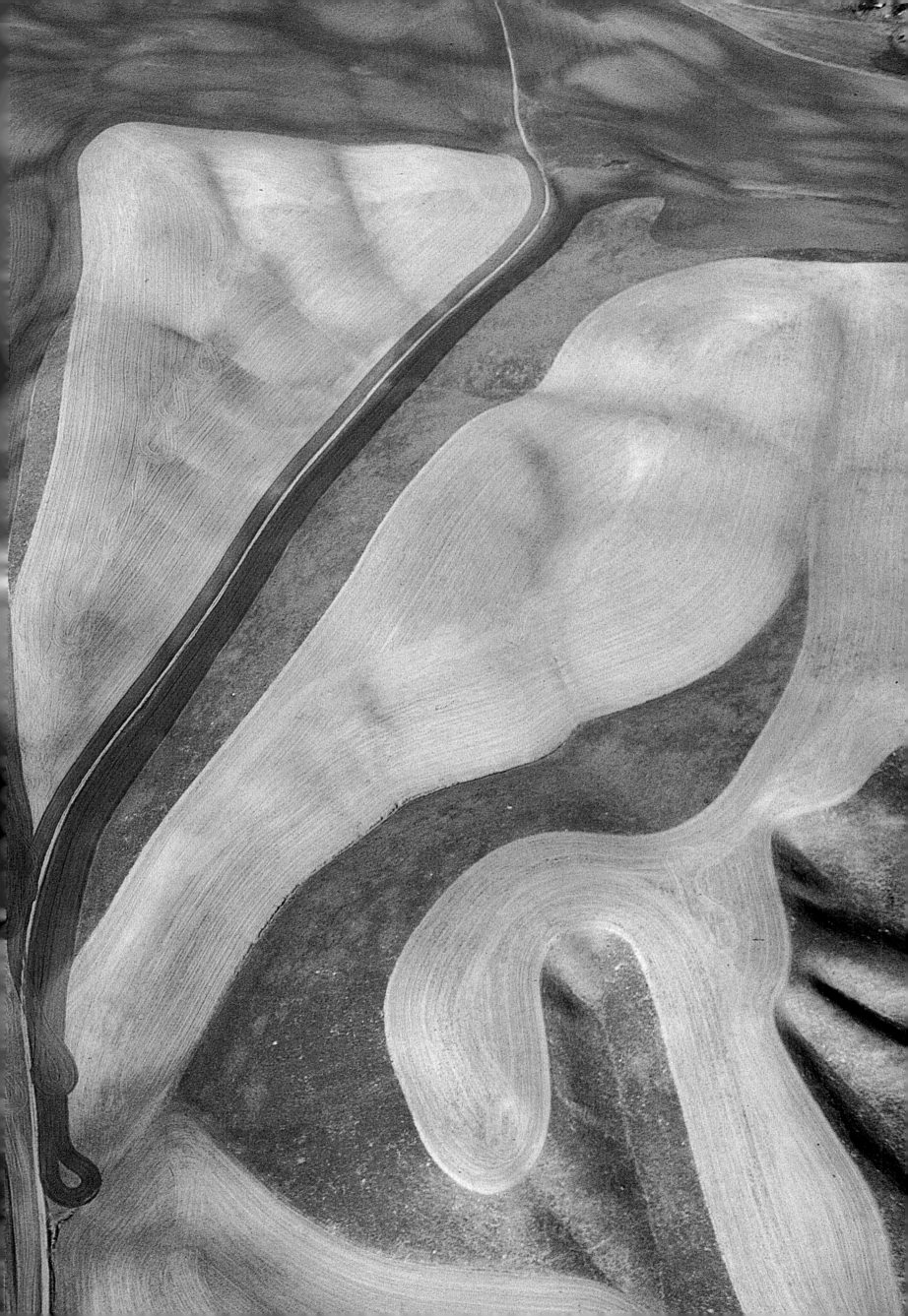

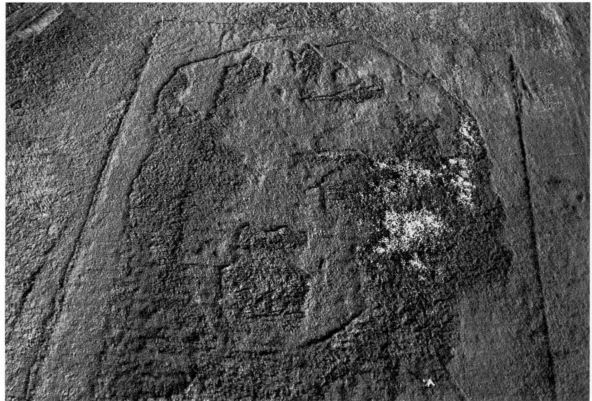

● *Above*

Icon of the age, "Andy Warhol's Marilyn" was fertilized in the grass beside a California freeway by artist Will Ashford, waving from the field. Photographer Georg Gerster repeated the image, à la Warhol, using false-color infrared film for the effect.

● *Right*

Another fascinating enigma, Leonardo da Vinci's "Mona Lisa," smiles down from the same slope above Interstate 680 near Alamo, California. Ashford applied nitrogen fertilizer to create both portraits through a lusher growth of grass.

246

GOD'S OWN LAND

"Those who labor in the earth are the chosen people of God, if ever He had a chosen people;" and America, Thomas Jefferson believed, was "the world's best hope . . . a chosen country." The dream of a land of opportunity existed long before America was discovered. The miracle was that America turned out to be not just another continent but a New World, a land commensurate with people's dreams. For some, America was an earthly Eden where God had shaken out His blessings as from a cornucopia. Naturalist William Bartram discovered paradise in North Carolina: "a vast expanse of green meadows and strawberry fields . . . flocks of turkeys strolling about . . . herds of deer prancing in the meads . . . innocent Cherokee virgins . . . disclosing their beauties to the fluttering breeze . . . too enticing for hearty young men long to continue idle spectators."

But for the stolid husbandmen who made up most of the populace, America was the Promised Land, fresh from the hand of God, awaiting the hand of God's Own People to transform the primordial wilderness into a garden. Hacking away at the virgin forest, the Puritans imposed a pastoral order on New England. Midwestern pioneers labored to break the virgin sod of the treeless prairies. "There was nothing but land," Willa Cather recalled, "not a country at all, but the material out of which countries are made." On the desolate shores of the Great Salt Lake, Mormon emigrants lost no time in setting to work to make God's desert bloom: "About two hours after our arrival," wrote one of His Saints, "we began to plough, and the same afternoon built a dam to irrigate the soil." Still questing in the 1930s, Dust Bowl refugees abandoned the drought-stricken heartland and took Route 66 west to the Promised Land of California, which they would help bring to fabulous fruition.

Primeval America filled early settlers with fear, awe, admiration – mixed emotions that evolved into conflicting desires to cherish the land yet exploit its resources, the age-old dilemma: to venerate or dominate nature. The agrarian advance often resembled an assault on enemy territory, but as expansion has played out, a sense of stewardship has developed on several levels. Grateful for the usufruct of God's Own Land, we recognize that land, as the stuff of creation, belongs ultimately to the creator. The land also constitutes a national patrimony for which we are trustees. The care farmers give to land they hope to pass on to their children is an expression of love. And we have a duty to the land itself, to use it wisely and well, keep it healthy and wholesome. Good farmers pass on more than land from generation to generation. They also transmit their intimate knowledge of the land and its products, traditions of hard work and good stewardship – a storehouse of practical skills and enduring values.

Most farmers come from a long line of people who have worked the land, but the links between generations and their common bond with the earth are now threatened. As mechanical power has replaced muscle power, the United States farm population has fallen from a peak of 32 million in the early 1900s to less than 5 million today. Farmers are older, with a median age of about fifty, and farm families are smaller. For every two adults living on farms, there is, on average, less than one child under fifteen years old. When farmers leave the land, or their sons and daughters do, a cultural continuum is broken, sundered by a cycle of losses – loss of opportunity, loss of an experienced farming community, loss of memories of the land and dreams for its future. If we are to keep the land's promise, we must husband our human resources as well as our soil and water. "In proportion as the ratio of farmers decreases is it important that we provide them the best of opportunities and encouragement: they must be better and better men," Liberty Hyde Bailey wrote in 1914 in *The Holy Earth*, a prophetic essay on the interdependence of human beings and the natural world. To keep the earth forever fruitful requires long-range vision; to preserve its beauty requires emotional commitment; to farm the land well requires a relationship with the earth that is at once practical, passionate, personal, pervasive, deep-rooted. "The proper caretaking of the earth lies not alone in maintaining its fertility or in safeguarding its products," wrote Bailey. "To put the best expression of any landscape into the consciousness of one's day's work is more to be desired than much riches . . . The lines of utility and efficiency ought also to be the lines of beauty." And he reminded us, "Man found the earth looking well. Humanity began in a garden."

Left: Wherever rainbows touch the earth, tradition says, there lie the most fertile soils. Amid the mountains of northern Utah, God points to the Bear River Valley, the state's most productive farming area.

Overleaf: Garden in the wilderness, a hilltop farm in the Blue Ridge Mountains near Winchester, Virginia, evokes, from a distance, a clearing carved out by a pioneer farmer in the 1700s. A closer look reveals a hobby farm created by twentieth-century city-dwellers longing for the country.

● *Previous pages*

BOUNTEOUS CORN harvest unfolds in southwestern Iowa, surveyed by a yellow sentinel of autumn, an isolated tree at the edge of an alfalfa field. Contoured terraces curb erosion of the deep loess soils, once held firmly in place by the matted root systems of the tallgrass prairie.

● *Pages 252-253*

THE ART of tilling puts poetry in the landscape and brings stability to the soils of southeastern Pennsylvania. A tradition of careful farming also imparts stability to families. In Lancaster County, where bonds with the land are strong, twenty-seven families have been cultivating their farms in an unbroken line for more than two hundred years.

INDEX

Entries refer to photograph captions.

PHOTOGRAPHER'S ACKNOWLEDGMENTS

I thank first and foremost the National Geographic Society of Washington, D.C., and in particular Wilbur E. Garrett, then the Editor of its Magazine. Their continued support, material and moral, has been a tremendous encouragement. In its September 1984 issue, the Magazine published a portfolio of my farmland aerials under the title "Patterns of Plenty."

My heartfelt thanks also go to the Soil Conservation Service of the U.S. Department of Agriculture and to the federal, state, and local offices of the Cooperative Extension System: Their numerous agents throughout the country efficiently assisted me by pointing out promising sites and, after the shoot, by helping to explain the what and the why in many a picture.

Moreover, I am grateful to the Soil Conservation Society of America and the American Farmland Trust, both headquartered in Washington, D.C. When invited to show some aerials of U.S. farmlands to the annual congress of the former organization's California chapter in 1986, I met with a heartening response that made the vague idea for a book coalesce into a firmly contoured project; I was especially spurred on by the warm reception given my images by Norman A. Berg, President of the Soil Conservation Society of America and the grand old man of soil conservation, and Robert Gray, the convention's keynote speaker, then with the American Farmland Trust. The Trust eventually became a partner in this publishing venture, and I could always count on helpful advice from its officers, notably its President, Ralph Grossi.

GEORG GERSTER

AUTHOR'S ACKNOWLEDGMENTS

My debts are many and great, to individuals and institutions too numerous to list.

For their patience and generosity, my sincere thanks go to all those who have served as my guides and teachers in this project, especially the dedicated and knowledgeable staffs of the Cooperative Extension System and of the Soil Conservation Service of the U.S. Department of Agriculture.

In the course of my research, I have consulted hundreds of books, articles, and other publications. For the information and insights I have gleaned from them, I gratefully acknowledge my debt to other writers and researchers as well as to associations, institutes, and foundations concerned with various facets of American agriculture.

Many have helped in the harvest, but for any errors that may appear — as tares among the wheat — I alone am responsible.

JOYCE DIAMANTI

PUBLISHER'S ACKNOWLEDGMENTS

For permission to reproduce text extracts, the publishers would like to thank the following. The epigraph on page 27 is from *Dream and Deed: The Story of Katharine Lee Bates* by Dorothy Burgess. Copyright 1952 by the University of Oklahoma Press. "Soils that melt like sugar" on page 150 is from *Competition for Land in the American South: Agriculture, Human Settlement, and the Environment* by Robert G. Healy 1985. Reprinted by permission of The Conservation Foundation. Quotations from Hamlin Garland on pages 27 and 231 are from *A Son of the Middle Border* by Hamlin Garland. Copyright 1917 by Hamlin Garland; copyright renewed 1945 by Mary I. Lord and Constance G. Williams. Reprinted by permission of Macmillan Publishing Company. "The tough vinelike roots" on pages 30–32 is quoted in *The Literary Guide to the United States* edited by Stewart Benedict. Published by Facts on File 1981. Reprinted by permission of Michael Friedman Publishing Group.

The extract from John Steinbeck on pages 34–35 is from *The Grapes of Wrath*. Copyright 1939, 1967 by John Steinbeck. Published by Viking Penguin 1939, 1986. The extract from Hugh Hammond Bennett on page 35 is from a speech given before the 78th Annual Meeting of the National Education Association, Milwaukee, Wisconsin, July 2, 1940, quoted in R. Neil Sampson, *Farmland or Wasteland: A Time to Choose*. Copyright 1981 by R. Neil Sampson. Published by Rodale Books 1981. William Gass's words on page 58 are from *In the Heart of the Heart of the Country and Other Stories*. Copyright by William H. Gass. Published by Harper & Row [n.d.]. The words of Johnny Appleseed on pages 101 and 104 are from Robert Price, *Johnny Appleseed: Man and Myth*. Copyright Robert Price. Published by Indiana University Press 1954. The quotations from Willa Cather on pages 125 and 249 are from *My Antonia*. Copyright 1918, 1926, 1946 by Willa Sibert Cather; copyright 1954 by Edith Lewis. Published by Houghton Mifflin Company [n.d.]. Aldo Leopold's words on page 164 are from *Round River* in *A Sand Country Almanac with Other Essays on Conservation from Round River*. Copyright Oxford University Press 1949, 1953, 1966. The quotation from Robert Frost on page 201 is from "The Gift Outright" in his *Poetry and Prose*. Copyright Holt, Rinehart and Winston 1972. The extract by Frank Norris on page 231 is from *The Octopus* in *Novels and Essays*. Published by the Library of America 1986. The extract from Walt Whitman on page 231 is from *The Complete Poetry and Prose of Walt Whitman . . . with an Introduction by Malcolm Cowley*. Published by Pellegrini & Cudahy 1948.